T0318229

Change in the Construction Industry

The UK construction industry is the sixth largest industry in the UK in terms of turnover. Efforts of successive Governments since the war to achieve an effective, viable, and efficient industry providing genuine value for money have recognised the vital part this industry plays in producing the country's infrastructure. Yet these efforts have, to a large extent, been unsuccessful due to the fragmented state of the industry, the adversarial nature of the industry's relationship with its clients and within its own structure, and the failure of government departments in the past to give a convincing policy lead to encourage investment and good management.

During the last decade, the industry has undergone an unprecedented period of self-examination, including input from most of the leaders of the major suppliers and clients, both private and public sector, as well as from leading ministers and opposition politicians, their civil servants and political advisers. Government and industry have attempted to work together to achieve political and structural change in the industry through collaborative action, and to bring about nothing less than a re-organisation of the way it undertakes its business.

Change in the Construction Industry details and documents, in an objective and factual way, what happened and the reasons for it, and offers an unbiased interpretation of the success or failure of the various initiatives that emerged, such as the Movement for Innovation, Rethinking Construction and Constructing Excellence. Drawing on personal accounts from politicians, civil servants, advisers, and industry leaders who were involved at the time, and who are willing to be quoted and identified, it provides invaluable source material for students of government/industry relations, for industry practitioners and clients, and for economic and social commentators.

David M. Adamson is Director of Estates at the University of Cambridge. During the Latham Review he was involved in the establishment of a comprehensive client representative body, the Construction Clients Forum (CCF) and was Deputy Chairman and Supervisory Board member. He was also on the Executive Committee of the Construction Industry Board and is currently on the Board of the Construction Skills/CITB. **Tony Pollington** was a career civil servant, serving in the old Ministry of Housing and Local Government, the Ministry of Transport, and the Department of the Environment and Property Services agency. He was the first Executive Secretary of the CCF and its successor, the Confederation of Construction Clients (CCC).

Routledge studies in business organisations and networks

Change in the Construction Industry

An account of the UK Construction Industry Reform Movement 1993–2003

David M. Adamson and Tony Pollington

Routledge
Taylor & Francis Group

LONDON AND NEW YORK

First published 2006
by Routledge
2 Park Square, Milton Park, Abingdon, Oxon OX14 4RN

Simultaneously published in the USA and Canada
by Routledge
711 Third Avenue, New York, NY 10017

*Routledge is an imprint of the Taylor & Francis Group, an
informa business*

First issued in paperback 2012

© 2006 David M. Adamson and Tony Pollington

Typeset in Times by Wearset Ltd, Boldon, Tyne and Wear

All rights reserved. No part of this book may be reprinted or
reproduced or utilised in any form or by any electronic,
mechanical, or other means, now known or hereafter
invented, including photocopying and recording, or in any
information storage or retrieval system, without permission in
writing from the publishers.

British Library Cataloguing in Publication Data
A catalogue record for this book is available from the British
Library

Library of Congress Cataloging in Publication Data
A catalog record for this book has been requested

ISBN13: 978-0-415-38599-2 (hbk)
ISBN13: 978-0-415-64647-5 (pbk)
ISBN13: 978-0-203-08800-5 (ebk)

Dedication

David Adamson dedicates this publication to his son Mark, with best wishes in furthering the aspirations for fairness and value that motivated the reformers of the British construction industry.

Tony Pollington dedicates it to the memory of his late brother Chris, who, as Deputy Secretary of the Conseil International du Bâtiment in Rotterdam, the Netherlands, contributed much to the industry internationally.

Contents

Figures

Tables

Acknowledgements

The authors acknowledge with gratitude the valuable contributions, both written and oral, which have been generously provided by those listed below, many of whom have also commented on early drafts of this account.

Michael G. Ankers Chief Executive of Construction Products Association

Nigel Chaldecott OBE Formerly Director of National Council of Building Materials Producers

Roger Courtney Formerly Director of Building Research Establishment

Alan Crane CBE Chairman of Rethinking Construction

Malcolm Dodds Director of Reading Construction Forum; Research Director of Collaborating for the Built Environment

Dr Frank Duffy CBE Formerly President of Royal Institute of British Architects

Sir Peter Gershon CBE Formerly Chief Executive of Office of Government Commerce

John Hobson Formerly Construction Director of Department of Environment

Stuart Humby Formerly Procurement Director of NatWest

Tony Jackson CBE Formerly Chairman of Construction Industry Board

Robert B. Jones Formerly Minister for Housing and Construction, Department of Environment

Richard B. Kauntze Chief Executive of the British Council of Offices

Professor Rudi Klein Chief Executive of the Specialist Engineering Contractors' Group

Sir Martin Laing CBE Honorary President of Laing

Sir Michael Latham DL Chairman of Construction Industry Training Board Construction Skills

Lord Levene of Portsoken KBE Chairman of Lloyds of London

William McKee CBE Formerly Director General of the British Property Federation; Chairman of Tilfen Land Limited

Anthony W. Merricks CBE General Manager of Balfour Beatty Specialist Holdings Division

Robin A. Nicholson CBE Senior Director of Edward Cullinan Architects

Richard G. Saxon Director of Building Design Partnership; Chairman of Collaborating for the Built Environment

Chris G. Sneath Chairman of the Plumbing and Heating Industry Alliance

Professor Ken W.J. Treadaway Formerly Director of Construction Round Table

Hugh W. Try CBE Formerly Chairman of Try Construction; Formerly Chairman of Construction Industry Training Board

Chris Vickers CBE Formerly Chairman of Construction Industry Board

Don Ward Chief Executive of Collaborating for the Built Environment

Phillip Ward Formerly Office of the Deputy Prime Minister

Graham Watts Chief Executive of the Construction Industry Council

Elizabeth A. Whatmore Construction Industry Directorate for Department of Trade and Industry

Geoffrey H. Wright Director of Project Management for Hammerson UK Properties plc

In addition to those mentioned above, many others have provided helpful contributions and comments, and our thanks are due to them, with our apologies for omission of specific reference to them.

Glossary

ACPS	Association of Construction Product Suppliers
AGILE	Agile Construction Initiative
BAA	British Airways Authority
BEC	Building Employers' Confederation
BMP	Building Materials Producers
BPF	British Property Federation
BRE	Building Research Establishment
BT	British Telecom
CABE	Commission for Architecture and the Built Environment
CBI	Confederation of British Industry
CBPP	Construction Best Practice Programme
CC	Construction Confederation
CCC	Confederation of Construction Clients
CCF	Construction Clients' Forum
CCSJC	Construction Contracts Standing Joint Committee
CIB	Construction Industry Board
CIC	Construction Industry Council
CIEC	Construction Industry Employers' Council
CIPS	Chartered Institute of Purchasing and Supply
CITB	Construction Industry Training Board
CLG	Construction Liaison Group
CPA	Construction Products Association
CPCG	Capital Projects Clients' Group
CPN	Construction Productivity Network
CRINE	Cost Reduction in the New Era
CRISP	Construction Research and Innovation Strategy Panel
CRT	Construction Round Table
CSCS	Construction Skills Certification Scheme
CUB	Construction Umbrella Bodies
CUP	Central Unit on Procurement
CVCP	Committee of Vice Chancellors and Principals

DBF	Design Build Foundation
DETR	Department of the Environment, Transport and the Regions
DfEE	Department for Education and Employment
DOE	Department of the Environment
DPM	Deputy Prime Minister
DTI	Department of Trade and Industry
DTLR	Department for Transport, Local Government and the Regions
ECA	Electrical Contractors' Association
ECI	European Construction Institute
EPSRC	Engineering and Physical Sciences Research Council
FCEC	Federation of Civil Engineering Contractors
GCCP	Government Construction Client Panel
GDP	Gross domestic product
HSE	Health and Safety Executive
ICE	Institution of Civil Engineers
JCT	Joint Contracts Tribunal
KPI	Key performance indicators
LGA	Local Government Association
M4I	Movement for Innovation
MCG	Major Contractors' Group
MOD	Ministry of Defence
NAO	National Audit Office
NCG	National Contractors' Group (of the Building Employers' Confederation)
NCW	National Construction Week
NHBC	National House-Building Council
NSCC	National Specialist Contractors' Council
OGC	Office of Government Commerce
PCIP	Productivity and Cost Improvement Panel
PFI	Private Finance Initiative
PPP	Public Private Partnerships
PSA	Property Services Agency
R&D	Research and development
RCF	Reading Construction Forum
RIBA	Royal Institute of British Architects
RICS	Royal Institute of Chartered Surveyors
SECG	Specialist Engineering Contractors' Group
SMMT	Society of Motor Manufacturers and Traders
SPATS	Scientific and Professional Administrative Training Schemes
TGWU	Transport and General Workers' Union
TUC	Trades Union Conference
UCATT	Union of Construction, Allied Trades and Technicians

Introduction

Construction is the sixth largest industry in the UK economy in terms of its proportion of total gross domestic product (GDP). The industry's output has been valued (2002) at approximately £65 billion annually, including design and management services, which represents well over 10 per cent of GDP.[1]

Given the size of the industry's output and its diversity it is almost inevitable that anybody engaged in defining the need for, and successfully implementing, capital investment, either in the private or the public sector, will at some time need to avail themselves of the industry's services.

Governments, as well as boards of directors, have long recognised that their ability to deliver the improved infrastructure necessary to meet society's aspirations and to provide a high level of productivity to compete in world markets, depends on having a technically and managerially efficient construction industry, responsive to customers' needs, receptive to innovation, and determined to provide value for money. In simple terms this means providing a well-designed quality product to meet clients' requirements, on time and within budget.

Unfortunately, the record of the UK industry in this regard in the 1960s, and 1970s, when the world was beginning to look for the rewards to be expected from the completion of the post-war reconstruction, was not good. Certainly there were examples of world-class building design in the UK, but these were in a minority. There were far too many examples of major cost and time overruns, poor quality design concentrating on minimising initial cost with no regard for longer term costs in use, and use of poor quality and inappropriate materials and components. Certainly the UK was widely regarded internationally as in the vanguard of resourcing research into the technical aspects of construction, but too often this was only academically based and the results abstruse and difficult to implement.

Governments since the war tried to remedy the situation. It was widely accepted that it was only the industry's clients that could lead the industry to define and adopt the cultural and structural changes required for it to achieve the necessary improved performance. Government-initiated attempts to review the state of the industry (Simon, Emmerson, and Banwell)[2] and hence to advance towards a collaborative, as opposed to adversarial, construction industry floundered mainly because the clients of the industry were not directly involved.

The industry too made serious efforts to identify its own weaknesses and to formulate policies to deal with them, notably the report by the major contractors group of the Building Employers Confederation, led by the chief executive of Tarmac Construction Ltd, Neville (later Sir Neville) Simms, and published as *Building Towards 2001* (see Appendix II).

The 1990s and the early years of the new century have, however, seen some remarkable changes in the way in which much of the industry and its customers now approach their business. Many now accept that an industry that bases its culture on mutual distrust and adversarialism between supplier and customer can never achieve the high levels of productivity and consumer satisfaction apparent in other consumer-based industries. Even if it is claimed by contractors and sub-contractors that the customers' motivating force in awarding work remains lowest cost, the fact is that many important invitations to tender stipulate partnering as the accepted method of defining and satisfying the requirement. It is now recognised as good client and supplier practice to invite all participants in an integrated supply team to contribute their own innovative ideas, utilising their particular expertise, towards the ultimate solution.

This change has come about through the extensive involvement, mostly on a voluntary basis, of many influential leaders from all sections of the industry, motivated by a common conviction of the necessity for change, encouraged over the last ten years by successive Governments which recognised that a better society, and their own political success, depended in large measure on an effective national construction sector. The movement that evolved from that conviction has seen comprehensive and insightful analysis of the industry's achievements and shortcomings (the Latham and Egan reports), the establishment of genuinely representative and comprehensive strategic pan-industry bodies (the Construction Industry Council (CIC), the Construction Industry Board (CIB), the Movement for Innovation (M4I) and the Strategic Forum for Construction), the first attempts to form a genuinely comprehensive single voice for the industry's clients (the Construction Clients' Forum (CCF) and then the Confederation of Construction Clients (CCC)), and attempts to integrate research and development strategies for the industry with the industry's own structural and managerial development Construction Research and Innovation Strategy Panel (CRISP).

There are numerous instances of initiatives run into the sand and abandoned, and of organisations established with great enthusiasm and hopes for the future which have, for various reasons, failed to fulfil their potential. Nobody could claim that the movement for reform from 1993–2003 has been universally successful. But the fact remains that the last decade has been stimulating, and certainly a pivotal phase in the development of relationships and mechanisms within and beyond the construction industry. Clearly major areas of the industry are better, and indeed more enjoyable and rewarding to work in. Now we appear to be in a period of sustained construction activity that will allow development and consolidation of the decade of reform, less buffeted by new initiatives.

The movement for reform in the UK construction industry during this period may well prove to have been one of the more successful examples of what happens when Government, irrespective of its particular political approach, seeks to work substantially together with an important industry in the economy, to effect major change, and adapt that industry to new and modern processes and procedures.

This decade of reform has required and inspired unprecedented wide-ranging, comprehensive and defined input over a long period from many expert sources inside and outside the industry. Because this is so valuable in itself, and because it carries with it lessons of immense value for those who may be involved in similar movements in the future, not necessarily only in construction, it is important that a dependable and authoritative record of what went on should be retained. This should be made accessible both to those who were involved at the time, and to students who wish to understand the background to, and reasons for, the changes that have been made in the industry. The first steps in this project were to find and safeguard key documents, and the personal views and recollections of people involved.

This record has two distinct objectives:

1   A factual, chronological account of the activities leading up to the Latham review, with the reasons for, and the actions following, that report, including the establishment of the Construction Industry Board; then the appointment of Sir John Egan and the subsequent progress towards the implementation of recommendations and integration, structural and operational change in the industry.

2   An analytical assessment of the resultant benefits of the actions undertaken, with an assessment of the actual outcomes, aimed at assisting those who may be involved in the strategic development or study of the industry in the future, in recognising the business case that can now be made for those in the industry who have not yet adopted the new culture.

Method of working

Inevitably, given the close involvement of both authors in the client movement during the period under review, much of what is recorded here reflects the clients' interpretation of what happened and the reasons for it. We recognise this, but remain convinced that we are justified in adopting this position, not least because it is the realisation that, without the determination of the clients to lead the industry in the improvements that have resulted from the reform movement, no real change would have been possible.

This is not to say that we have consciously downplayed the influence and achievements of the other sectors of the industry. We have tried to reflect accurately and in depth the contributions and observations given to us through personal interviews and in writing by the many participants in the movement, and whose names are acknowledged, with our sincere thanks and appreciation, on pages xi and xii.

Where those who have contributed material have subsequently suggested amendments or corrections to early drafts sent to them for comment, we have attempted to include these wherever possible. This record is not, however, in the nature of a Government or official industry report, but is the attempt of two of those intimately concerned to faithfully describe their view of what actually happened, seeking clarification and correction where possible from others concerned. The inevitable faults and inaccuracies are the responsibility of the authors themselves.

Notes

1 Professor David Pearce – nCRISP 2002.
2 *Report of the Central Council for Works and Buildings chaired by Sir Ernest Simon. The Placing and Management of Building Contracts*, HMSO 1944; *Survey of Problems before the Construction Industries*, H.C. Emmerson, HMSO 1962; *Report of the Committee on the Placing and Management of Contracts for Building and Civil Engineering Work chaired by Sir Harold Banwell*, HMSO 1964.

1 The state of the industry prior to the 1990s

Why reform was considered necessary. The background to the decision to have an independent review of the industry. What were the matters which showed the need for reform? Government concern. Decision to appoint Sir Michael Latham to conduct the review. Objectives of the review. How effectively did these connect with the modalities and practicalities of reform?

The decision of the Conservative Government to mount a 'Joint Review of Procurement and Contractual Arrangements in the United Kingdom construction industry', announced in the House of Commons on 5 July 1993 by the Under Secretary of State at the Department of the Environment, Tony Baldry MP, was a response to internal and external criticism of the industry over a long period. There was serious concern throughout the UK economy that the UK construction industry was not well placed to meet the challenges facing it from private and public clients, who looked to the industry to provide the construction infrastructure necessary for them to succeed in their business objectives. There was conjecture that the construction industry might follow the example of the British car, motorcycle, and white goods industries into foreign ownership.

The industry had suffered badly during the recession of the early 1990s. Output had declined 39 per cent between 1990–1993; half a million jobs had been lost in the industry between 1989 and 1993, and 35,000 small businesses had disappeared through insolvency. Training expenditure, vital to provide, maintain and improve the industry's skills, had reduced by 50 per cent in the same period.[1] The result was an industry with low productivity, and low profitability compared to other industrial sectors of the economy, leading in turn to a poor image with potential investors, under-capitalised, and with little motivation to re-structure itself in line with changing aspirations.

The problems of the industry were not, however, merely the result of sharp, but temporary, recession. They went much deeper and reflected an outdated attitude on behalf of both suppliers and customers towards an industry essentially concerned with providing services.

The industry was predominantly adversarial in its approach to its customers. This was reflected not only in an unwillingness to take risks, but also in a reliance on legalistic and tightly drawn contracting procedures, and recourse to law as the tried and tested method of resolving disputes and defending profitability. Unlike much of manufacturing industry, the supply team was fragmented and hierarchical, with little opportunity for specialists to contribute their expertise to the customer's benefit. The results were a reluctance to introduce innovative solutions to customers' requirements, an acceptance of the 'status quo' as a means of protecting professional and sectional interests, and a product too often representing the lowest acceptable levels of quality and design, instead of seizing the opportunity to provide world-class solutions, to delight the customer and enhance the prestige and locus of the UK internationally.

The supply side of the industry did not necessarily see it like this, however. Irrespective of whether they were so-called main contractors, sub-contractors (or 'specialist' contractors), 'professionals' or materials and components suppliers, every sector of the supply side of the industry was obsessed with volume. The trades unions shared this obsession. Throughout the 1970s and 1980s the burden of the industry's message to successive Governments was that the level of demand on the industry should be consistently increased, or at the very least held constant in real terms. And the industry had a point. As the sixth largest industry in the UK in terms of proportion of GDP, disposing of some £60 billion in terms of demand, nearly 60 per cent of which was directly controlled by Government as client, the industry was regularly used by the Government of the day as an economic regulator, with demand being turned on and off as the need for expansion or retrenchment fluctuated for economic or political reasons.

The fragmentation of the industry was highlighted by the conflict between main contractors and sub-contractors, stemming to some extent from the inadequate capitalisation of the majority of the large national contractors. The sub-contractors blamed this for a perceived tendency on the part of main contractors for delays in the payment process, particularly to specialists and sub-contractors. As far back as April 1991, the specialist engineering sector had asked the Government to institute an inquiry into contractual relationships within the industry; a request backed up by many letters from specialist contractors to ministers and MPs complaining about their alleged treatment at the hands of main contractors.

In late 1987, the National Contractors' Group (NCG) of the Building Employers' Confederation commissioned the report *Building Britain 2001* from the Centre for Strategic Studies in Construction in the Department of Construction Management and Engineering at the University of Reading. This report examined many areas of the construction industry, including market operations, management, research and development, education and training. Numerous recommendations were made and a further report, *Investing in Building 2001*, was commissioned and published subsequently in autumn 1989.

This second report focused in greater detail on the issues that were felt to be particularly important, namely image, organisation and structure, research and development, and education and training. The foreword to the report was written by Prime Minister Margaret Thatcher and effectively endorsed a series of actions. Shortly after publication, the Prime Minister requested that the Government be kept informed of progress, by reporting back to the Secretary of State for the Environment within 12 months (see Figure 1.1).

The NCG set up four task forces to cover the four main areas of concern described above and produced a comprehensive statement on the industry with detailed proposals for improvement under the generic title of *Building Towards 2001*.[2] The key points of the report, which helped to formulate the subsequent approach adopted by Sir Michael Latham (see Chapter 2), are summarised in Appendix II.

Prior to 1987/1988 the attitude of the industry, insofar as it was expressed collectively, was represented by the so-called Group of Eight, made up of the Royal Institute of British Architects (RIBA), Royal Institute of Chartered Surveyors (RICS), Institution of Civil Engineers (ICE), Building Employers' Confederation (BEC), Federation of Civil Engineering Contractors (FCEC), Building Materials Producers (BMP) and the Transport and General Workers' Union (TGWU) and Union of Construction, Allied Trades and Technicians (UCATT) from the trades union side. Virtually all of their limited

10 DOWNING STREET

LONDON SW1A 2AA

THE PRIME MINISTER

2 November 1989

Dear Mr Simms

As you know, I provided the foreword for the recently published report 'Investing in Building 2001' which was sponsored by the National Contractors' Group. Your Group will no doubt be keeping a close eye on the progress made by industry in responding to the action plan set out in the report, which I commended then. I would be very grateful if you would keep the Government informed of the progress being made. Perhaps you might let Chris Patten at the Department of the Environment know in twelve months time what steps have been taken.

Yours sincerely

Margaret Thatcher

Neville Simms, Esq.

Figure 1.1 Letter from Prime Minister Thatcher to Neville Simms asking to be kept informed of progress made by the industry in responding to *Investing in Building 2001*.

contacts with Government were concerned with the volume of demand on the industry, with continuous pleas to Government (Conservative) for more public investment and more in the way of 'handouts' through fiscal and taxation advantages. This went down badly with a Conservative Government facing a need to cut public expenditure, and essentially disengage from industry generally. With Owen Luder (president of the RIBA) as its chairman, the Group of Eight unwisely sought and obtained an hour's interview with Prime Minister Thatcher at 10 Downing Street. One of those present recalls that after having listened in silence to the industry representatives' pleas for help, the Prime Minister lost patience and sent the group away with a flea in their collective ear, emphasising that the industry's fate was in its own hands. It was made clear that if the UK construction industry was incapable of performing in a modern deregulated economy, the Government, and the public sector generally, would obtain its construction requirements from overseas sources. (Another attendee, however, recalls that it was the Group of Eight that had to listen in silence! Either way, the absence of dialogue is illustrative of the mood of the times.)

The result was the sidelining of the Group of Eight, and a proposal, originating with Ted Happold and other professional institute presidents, for a 'fraternal' council incorporating the main institutes and leading main contractors. This led to the formation of the Building Industry Council, later to become the Construction Industry Council (CIC) to complement the Construction Industry Employers' Council (CIEC), established by the employers with trades union participation. Ian Dixon, then chairman of Willmott Dixon and later to be knighted for his services to politics and the industry, followed Sir Edmund 'Ted' Happold as chairman of CIC. Over this period politicians as senior as Michael Heseltine were involved in discussions as to whether Government should lead a change in the industry, or whether improvement should be left to 'market forces'.

Although Ted Happold invited the employers' organisations, the BEC, FCEC, and BMP to join the Building Industries Council, they preferred to establish their own 'umbrella' body, the Construction Industry Employers' Council. Relations between the two organisations were, according to the CIC's chief executive, Graham Watts, who served the Council throughout this time and up to the present day, reserved at first, but both Sir Brian Hill and Ian Dixon worked hard to improve the atmosphere.

In 1991, Sir Brian Hill (chairman of Higgs and Hill) encouraged collaboration between CIEC and CIC, and issued a joint manifesto aimed at identifying the most important areas of industry activity which required improvement if the industry was to maintain, and eventually increase, its markets. In 1992, Secretary of State for the Environment Michael Howard, and Construction Minister, Sir George Young, attended the main CIEC industry dinner, and were subjected to some fairly 'robust' criticism (including the hurling of bread rolls) about the Conservative Government's alleged failure to support the industry's call for a major review of its shortcomings. Two main strands of activity were identified:

* the industry's relationship with Government; and,
* the industry's internal structure and relationships.

The 1992 election then intervened and, rather to the surprise of most industrialists, the Conservatives under John Major were returned, albeit with a small, and increasingly precarious, majority. Within a week, Secretary of State Michael Howard had convened a meeting to consider a major industry review. Tony Baldry became Junior Construction Minister, working to Sir George Young, and he identified the industry's relationship with Government as the area on which to concentrate. He initiated fortnightly industry briefings, attended by departmental officials as well as ministers, which were well regarded by those participating in them, and these regular briefings provided and sustained an on-going momentum for a basic and comprehensive review.

Tony Merricks, who came from the contracting side of the industry, and who was to become in due course one of the deputy chairmen of the Construction Industry Board (CIB), speaks highly of the success of Tony Baldry as minister in helping to bring the main contractors and specialist sub-contractors into a more productive relationship, and to the joint recognition that, unless the industry changed, there were 'no decent career prospects unless you were a lawyer'. Although Tony Merricks believes that in the long run the industry failed to take full advantage of the opportunity thus offered, this influence of the Government at the time helped to consolidate the efforts of the specialist sub-contractors to get the industry to speak with one voice.

This attempt by the Government of the day to encourage the industry to develop sound relationships was well appreciated by the industry. Indeed, Tony Jackson, then closely involved with the industry's affairs in his capacity as director of Blue Circle, developers of the major Thames crossing sites, and as chairman of the National

Council of Building Materials Producers, has said that the industry enjoyed the best relationships with Government at this time in its history.

Much of the credit for this must go to the civil servants then in the DOE, responsible for advising ministers. In the period immediately prior to the decision to initiate the review of the industry, four different officials had been appointed as head of the Department's Construction Directorate in four years. The resulting inability to forge close links with the Government through permanent officials contributed to the industry's failure adequately to put its case across to policy makers, an important matter.

With the growing conviction within the Government and industry for the need for a radical review, the DOE appointed Phillip Ward as head of its Construction Directorate in 1992. This proved an inspired choice, with Phillip rapidly establishing an atmosphere of respect and trust, and a conviction within the industry that its representations were listened to, and that it had ready access to the Government on matters of concern to it. This is not to say that Phillip Ward could be regarded in any way as a soft touch, and he was not, by his own account (interview 9 July 2003), prepared to advise cushioning the industry against the prevailing policy of the Treasury economists, which was that suppliers within the UK's domestic economy should not be protected against competition. Nonetheless, despite the industry's previous perception that it had always done better in terms of workload under Labour Governments than under the Conservatives, the industry accepted that the Government was supportive of its continued efforts to obtain work abroad, and that it wanted it to succeed.

Within a developing atmosphere of unhappiness with and within the industry, and the shifting relationships with it, the proposal for a full review of the industry, and then the choice of Sir Michael Latham to undertake it, gained increasing support, although it was by no means plain sailing. There was considerable jockeying for position among the various industry sectors, with the main contractors being particularly concerned that any industry examination which involved the sub-contracting interests would lead to loss of power for the main contractors. The professional bodies, too, the RIBA, RICS, ICE, and the other engineering institutes, saw a comprehensive review as likely to lead to the surrender of professionals' claims to lead the building team in the traditional sense; it was predominantly through the efforts of the Construction Industry Council that the respective professions agreed to subordinate particular and potentially conflicting self-interests to the advantage of the industry generally. Phillip Ward, in early consultations with Sir Michael Latham before the latter was appointed, encouraged the various representative bodies in the industry to focus on the importance to them of shared objectives for the industry as a whole, and this paved the way for the acceptance of Sir Michael Latham's appointment. Several others were suggested for the study before Sir Michael was asked, but each was unacceptable (or thought he would be unacceptable) to one or more sector groups. The sub-contractors in particular had initial reservations about the appointment of Sir Michael Latham, stressing his previous involvement as a consultant to the Building Employers' Council, and therefore possibly unsympathetic to their position. They favoured the appointment of Patrick Nicholls MP as joint reviewer with Michael Latham, bearing in mind that Patrick Nicholls had acted previously as consultant to the National Specialist Contractors' Council (NSCC). However, following the Electrical Contractors' Association (ECA) dinner in February 1993, at which Michael Latham deputised as speaker at short notice for Sir George Young, ECA and BEC leaders approached him to see if he would be interested in principle in taking on the task. In retrospect, it is clear that in Latham the best choice was made, by virtue both of his universal acceptability and also by virtue of the intellectual rigour he applied: the rigour and clarity of the historian (he had gained a first in history at King's College, Cambridge) was coupled with the experience from 18 years' service as a Member of Parliament.

Notes

1 *Constructing the Team: Report by Sir Michael Latham*, HMSO, July 1994.
2 *Building Towards 2001.* A report by the National Contractors' Group of the Building Employers' Confederation and Reading University's Centre for Strategic Studies in Construction. Published for BEC by *Building Magazine* 1989.

2   The Latham review

Establishment: appointment of assessors: interim report *Trust and Money* published December 1993. Final report *Constructing the Team* published July 1994. The report's conclusions and recommendations: assessment of their validity. Validity assessed against objectives.

Once Sir Michael Latham was formally appointed, and the review announced in Parliament, major effort was devoted to ensuring after many months that all sectors of the industry, irrespective of their early doubts about how the review might impinge on their particular sectoral interests, would contribute to it. Because the industry at large recognised the commitment of the DOE to a genuine examination, without overt political overtones, and because the industry had confidence in Sir Michael Latham's knowledge of the industry (he had been a housing advisor in the Wilson Government 1965–1970, and later in the Conservative administration under Sir Edward Heath, and had also been a director of the House Builders' Federation), the terms of reference of the review (below) were accepted, albeit after a considerable length of time while the various sectoral interests sought to protect their particular interests.

Terms of Reference for the Review

To consider:

- current procurement and contractual arrangements; and
- current roles, responsibilities and performance of the participants, including the client.

with particular regard to:

- the processes by which clients' requirements are established and presented;
- methods of procurement;
- responsibility for the production, management and development of design;
- organisation and management of the construction processes;
- contractual issues and methods of dispute resolution; and

in doing so to take into account:

- the structure of the industry;
- the importance of fair and transparent competition;
- the desirability of a fair balance between the interests of, and the risks borne by, the client and the various parties involved in a project;
- the requirements of public accountability, value for money and EC legislation as regards public sector contracts;
- the importance of encouraging enterprise, the development of a skilled labour force and investment in improving quality and efficiency;
- current developments in law;
- relevant comparisons with the structure and performance of the construction industry in other countries;

with the objectives of making recommendations to Government, the construction industry and its clients regarding reform to reduce conflict and litigation and encourage the industry's productivity and competitiveness.

That so many of the objectives were achieved was due in great measure to the skills of Sir Michael himself, and of the senior civil servant heading the Department of the Environment's Construction Directorate, Phillip Ward, in convincing the various sectors of the industry that they all stood to win rather than lose from the review's identification of the priorities for improvement.

Nonetheless, there were difficulties in involving some parts of the industry in the early stages. Several of those involved at the time, including Sir Martin Laing in the CIEC, have confirmed that the main contractors in general originally opposed the whole concept of the review, but went along with it, faced with the clear indication of the rest of the industry's enthusiasm, and the recognition that, enjoying as it did strong support from the Government, the review would be undertaken with or without them.

The mode of working adopted by Sir Michael and the review's secretary, Deborah Bronnert, seconded by DOE, was to appoint assessors (see page 1 of the report) representing the four industry organisations (CIEC, CIC, NSCC, and the Specialist Engineering Contractors' Group (SECG)) together with representatives of the DOE and two organisations which claimed to represent the industry's clients, the British Property Federation (BPF) and the Chartered Institute of Purchasing and Supply (CIPS). This method of proceeding undoubtedly worked well, in that it placed responsibility for obtaining comprehensive responses to the issues listed in the terms of reference firmly with the industry itself, rather than having them subjected to formal interrogation from a Government-appointed team as in a formal Government inquiry.

One of the main reasons why the Latham review appeared to receive greater support from the industry than the several previous inquiries, such as Simon and Banwell, was the perception that, for the first time, the clients were to be formally involved in the development of a strategy for the industry. The consumer movement had spread to construction. The difficulty here was that no comprehensive client body existed which could be called upon to provide a genuinely expert, coherent and comprehensive voice.

The BPF was, of course, the most important representative trade association for the property development sector, most of whose members were involved in buying construction not for their own occupation but for letting or selling on to other users. As such, however, they were not universally regarded as totally representative of the industry's customers, although clearly the size of their demand on the industry's resources (approximately £14bn in 1990) identified them as a very influential group. Some of the larger BPF member firms (Hammersons, Slough Estates, and Metropolitan Estates, for example) had much expertise to deploy, often through their own construction departments, staffed by construction professionals. From an initial position of being implacably opposed to the idea of an industry-wide review, the Federation's position under its new director-general, Will McKee, softened its line. As a result, the property development sector exercised a great deal of influence in the identification of what issues the review should cover, in its actual establishment, and in its early implementation.

Will McKee, has described the involvement of the Federation from the earliest days. Frequent contacts between the BPF and the

DOE, which was the property sector's contact with central Government, had identified on-going concerns by the Federation's members about the relatively high incidence of cost and time overruns on projects financed by the property developers as investors, as well as unresolved, detailed matters which, in the BPF members' opinion, adversely affected their commercial interests. The BPF members were particularly concerned about the need for collateral warranties (their perception that the supply side was excessively cautious in its acceptance of risk) and long-established ill feeling on the subject of retention. The Federation, however, tried to take a more proactive and positive approach, seeing these admittedly unresolved detailed matters as evidence of the wider malaise affecting the industry. In discussions with the Minister at the time, Tony Baldry MP, the Federation pressed very hard for any review to include full representation of the industry's clients, despite considerable scepticism by some of the Federation's members who retained a great deal of dissatisfaction with the perceived failure of earlier reviews. The president of the Joint Contracts Tribunal at the time, Roger Squire, then director and deputy chief executive of the London Docklands Development Corporation, and the BPF director-general, presented the Department with a list of client concerns, leading to the conclusion that the prevailing theme of the review should be the examination of the options for effective procurement of construction and how these impinged on clients' ability to achieve these business objectives.

The BPF has maintained throughout that its motivation in pressing for the review was not dictated wholly by self-interest, but by the conviction that the economy generally, and thus its members' commercial activities, could only be enhanced by improvement in the industry's efficiency. For this reason, according to its director-general, the Federation continued to pursue other campaigns outside the remit of the review, for example a fundamental review of the structure and working of the Joint Contracts Tribunal whose family of construction contract forms were widely used, and strongly criticised, by BPF members, and other clients. (Nearly all its drafting was done by 'the supply side' and so it reflected the interests of those unwilling to ease the client's liability for risks.)

The involvement of CIPS was anticipated to provide access to a wide selection of client organisations, since the Institute's members, all individuals, exercised the procurement responsibility within public and private sector organisations. Certainly the Institute's formally appointed assessor, Frank Griffiths, a former president, and the team set up within the Institute under Stuart Humby, another of its former presidents, to support him, had wide experience in the procurement of construction, particularly in the civil engineering and process engineering fields, notably with regard to the commissioning of very large schemes for the national utilities. This experience was well reflected in the interim report of the Latham review, which appeared in 1994 under the title *Trust and Money*, and which set the parameters for the review's final report. The CIPS contribution to this, which was submitted to the review as 'Productivity and Costs', strongly influenced the content and the style of the final report, and the Institute's proposals for establishing consultative machinery within which the industry could discuss and formulate its strategy was reproduced as a separate appendix (Appendix VI) to Sir Michael's final report.

Nonetheless, there must remain some concern that the clients' input to the review should have been more truly representative of the wide spectrum of the industry's customers. In particular it proved difficult to obtain high-level input to the review from important sectors such as retailing, leisure and manufacturing industry. Sainsbury's, who, with some other major retailers such as Marks and Spencer, had participated in private discussions within loosely organised pan-industry forums such as the Construction Round Table, and who possessed considerable expertise in relation to the procurement of standard solutions for particular business requirements, contributed valuable material. However, in common with other commercial client sectors, they were unable to devote time and resources to marshalling truly comprehensive coverage of the interests of their sectors, and to a certain extent were inhibited from doing so because of the risk to commercial confidentiality concerning their competitors.

This perception of the background to the establishment of the Latham review, and of the efforts undertaken to obtain the support and involvement of the different sectors of the industry, despite their differences, is confirmed by Sir Michael Latham's own recollections. In an interview in June 2003 he gave the following description of what happened.

The need for a fundamental review of the UK construction industry first surfaced at the biennial meeting of the Building Employers' Confederation in November 1991. Held in the National Convention Centre in Birmingham, the idea was first mooted of a 'dispute-free' industry, largely because of exasperation on the part of the employers with the dispute-ridden and litigation-happy approach of both clients and suppliers. Some 6 months later, the 1992 annual general meeting of the National Contractors Group of the BEC saw Sir Neville Simms, then chief executive of Tarmac (later to become Carillion plc), call for an industry-wide review, bearing in mind not just dissatisfaction expressed by client and supply sides with the confrontational approach to business, but also the dawning realisation that performance was deteriorating in relation to both national and international indicators.

As it happened, this meeting was attended by the then Conservative Housing and Construction Minister, Sir George Young MP, who welcomed the idea, and promised full Government support.

Before this could be implemented, a general election was called, and the Conservatives returned, albeit with a very much reduced majority. Sir George Young was confirmed as continuing as Minister with responsibility in the Commons for construction and one of the first actions as a Minister in the new Government was to ring up the BEC and press for them to lead the review.

The terms of reference that then emerged were essentially defensive in tone, with every section of the supply side rigorously guarding their backs. Evident from the start was a basic dispute between the main (traditional) contractors, and the specialist sub-contractors. This dispute, which in reality involved a realisation that, with increasing and varied skills and specialisation, a hierarchical supply chain, in which sub-contractors answered to an all-powerful main contractor, was of decreasing validity in the modern world, crystallised into deeply held opposition on the part of the specialist contractors to the concept of 'pay when paid', and other perceived tendering abuses.

Despite this, however, the BEC was adamant that they did not want the kind of official, hierarchically structured and Government-directed review personified by such earlier attempts as Simon and Banwell. They strongly favoured a one-person review, adequately supported, preferably by the industry's supply side and enjoying Government backing without Government control. At that time (December 1992) there was no immediately recognised candidate to undertake the review, who would be accepted by all sectors. Within (Conservative) Government circles several names were mentioned but were not strongly supported by the industry, and at least one was in fact vetoed by Sir George Young, the Minister, on the grounds that he would not write the report himself, which was what the combined industry really wanted. Little thought had been given at this time towards the involvement of the clients of the industry, whose dissatisfaction with the quality of the service they had experienced was largely responsible for the call for action to improve the industry. It was as a result of the intervention of the then director general of the British Property Federation, Will McKee, that a group of representatives of clients and suppliers, composed of representatives of the British Property Federation, the Chartered Institute of Purchasing and Supply, the Construction Industry Council, the Specialist Engineering Contractors' Group, the National Sub-Contractors' Council, and the Construction Industry Employers' Council, gave coherent voice to the growing realisation that no attempt to improve the industry stood any chance of success unless it was led, or at least strongly influenced, by the industry's customers. As a result of this conviction, Ian Deslandes, then Director of the BEC, at the February 1993 Annual Dinner of the Electrical Contractors' Association, raised the issue directly with me.

After being assured that the sub-contractors would in fact collaborate, despite their antagonism to, and distrust of, the main contractor, I agreed to undertake the review he demanded, and was given Departmental administrative support and complete autonomy. Following initial soundings among the constituent parts of the industry, and among Departmental officials, I concluded that Ministers' original remit, to investigate means of mitigating the adverse effects of the traditional adversarial approach to construction contracting, was too narrow in scope. I accepted the Department's view that the review should be all embracing, involving not just traditional construction and its modern developments such as pre-fabrication and off-site manufacturing, but also civil and process engineering. I accepted the view that the study should encompass the situation throughout the UK, including Northern Ireland.

The final report, under the title of *Constructing the Team*, was presented to the Government and published in July 1994. Because Parliament was in recess, its publication was not notified to Parliament until the new session, by which time action had already started on implementing its recommendations.

The report's recommendations are reproduced in Appendix I and fall into the following categories.

Recommendations 1–5 were aimed at identifying the client as the leader in the process and providing organisational and process structures to allow this to be done.

Then followed a series of recommendations (6–8) concerned with aligning the design function with the interests of the client.

Recommendations 9–12 were aimed at encouraging clients' understanding and use of contract documentation, and the development of new, more modern and client-friendly collaborative forms of contract.

Major work on developing codes of practice for selection and tendering procedures (recommendations 13–19) was intended to promote a more integrated and partnering-based approach to project management, and the establishment of truly integrated supply chains, as already established in other productive sectors of the national economy.

Determining the level and quality of the industry's performance, and facilitating the availability of more skilled resources through education, training and research (recommendations 20–24) were intended to drive clients and suppliers towards recognition that measurement of performance is the prerequisite for deciding whether or not the industry is improving.

Removal of unfair, one-sided or obscure contract conditions, and the establishment of fair methods of payment, positive and proactive dispute resolution procedures and safeguards against insolvency (recommendations 25–27) were intended to pave the way for clients and the supply side to move towards an attitude based on equitable sharing of risks and rewards. The concept of 'fairness', central in the report, was not hitherto recognised as one of the fundamentals in the industry: it was later to be challenged by lawyers. There were specific recommendations (recommendations 28–29) for legislation to deal with responsibility for, and insurance against, liability, including latent defects insurance. Finally, specific proposals were made (recommendation 30) for mechanisms designed to deliver the required improvements within a defined timescale.

It is a reflection of the comprehensive nature of the investigation, and the skilful way in which it was undertaken, that these numerous recommendations, many requiring a radical rethink by the industry's own vested interests, were almost universally accepted and received with great public acclaim. Only the recommendation for mandatory trust funds (recommendation 27) did not proceed because many (though not all) of the influential clients were unhappy with it and the main contractors and sub-contractors could not agree amongst themselves on how trust funds should be expected to work in practice.

3   Action resulting from Latham

The Government response; acceptance by Treasury and Central Unit on Purchasing (later Office of Government Commerce (OGC)). The National Conference 1994; key agreements to:

- introduce Construction Bill;
- establish industry-wide strategic body, the Construction Industry Board (CIB);
- give predominance to clients.

As might be expected, the report and recommendations of Sir Michael Latham's review, published as *Constructing the Team*, were welcomed by the Department of the Environment and by the Government generally. It enjoyed a very good press from the technical and professional construction industry publications, particularly *Building*, *Contracts Journal* and the *Architect's Journal*. The official publications of the professional institutes, RIBA, RICS, and ICE were more guarded, including in their coverage concerns about qualified professionals losing their independence in relation to main contractors and hence their ability, in their eyes, objectively to represent the interests of clients. It should, however, be noted that the president of the RIBA, Frank Duffy, later proved very supportive of the approach adopted and effective in advancing the involvement of designers (although that influence dwindled after his presidency). On the other hand, there remains ground for criticism that the review was perceived as very much London-based, and there seemed for a long time to be little recognition of its objectives and conclusions in other regions of the UK.

This situation partly reflected the difficult relationships between the professional bodies and the representative organisations of the contractors. The latter rapidly identified the Latham review as signing the death warrant for the old dispensation, whereby what was known as the 'main contractor', who inevitably was the principal in the traditional forms of contract with the client, controlled the supply side and their methods of delivery to the client. The Secretary of State in the Department of the Environment at the time of the commissioning of the review, Michael Howard, and his ministers with overall responsibility for the Governments' contacts with the industry, Sir George Young and Tony Baldry, had identified the industry's poor relationship with Government, and worse relationships with its own constituent parts, as the major weaknesses requiring solution if there was to be any hope of lasting improvement in performance. The Latham review overwhelmingly endorsed this, but the review itself could only expose the scene and offer possible ways ahead, implementation of which had to be in the industry's own hands.

If there had been a weakness in the way Sir Michael conducted his review, it had been his perceived reluctance to involve his assessors in the development of his recommendations, once he had received their detailed inputs relating to their sectors' particular interest. This appears to have been a conscious decision, aimed at convincing all the factions that his approach was genuinely open-handed and disinterested, but it left contractors, sub-contractors and professionals openly jockeying for position in the period before the review's publication. It did, however, reflect the wish of at least one representative body, the BPF, that the assessors should not be perceived as themselves committed to the recommendations. The philosophy was that the report should be one person's report, not that of a committee like Banwell. Graham Watts, chief executive of the Construction Industry Council (CIC), which attempted to bring together the concerns of the professional institutes, which were predominantly involved in design, has indicated that the CIC assessors, Robin Nicholson from the RIBA and Robin Wilson from the ICE, had a particularly difficult time in their relationships with the contractors' representative bodies. There seems to be a widely held view among the professional institutes that, had it been possible successfully to achieve such an objective, the Federation of Civil Engineering Contractors (FCEC) and the Major Contractors' Group (MCG), within the Construction Industry Employers' Council (CIEC), would have been happy to have seen *Constructing the Team* join Simons and Banwell on the shelf. The Major Contractors' Group, formed to represent the particular concerns of the 'big players' on the contracting side, was particularly identified as being less than enthusiastic about the review. However, Sir Martin Laing who was particularly active on behalf of the major contractors and worked closely with the MCG's director, Jennie Price, feels that their approach was dictated more by concerns about the effect on their companies' cash flow of giving a stronger position in the process to the sub-contractors, than by opposition to the recommendations of the review as a whole.

Perhaps as a result of an informal alliance between the CIC and the main sub-contractors' and specialists' groupings, the Specialist Engineering Contractors' Group, including the Electrical Contractors' Association and the National Specialist Contractors' Council, the main contractors seem to have been out-manoeuvred. With the professional institutes having been influenced by Ted Happold, and others making inputs to the review, to come together, the CIC found itself strongly placed to promote the idea of a major national conference to launch the review and its recommendations, and to influence the industry to adopt the collaborative approach to improving itself which the report advocated. What gave a considerable boost to this early idea was the indication, through Phillip Ward's close contacts with the various parts of the industry, that the Government was sympathetic to the important recommendation by Sir Michael that, in order to show its commitment to achieving change, the Government should seek to become a 'best practice' client.

Against this background, the Prime Minister, John Major appointed Robert Jones as minister with responsibility for construction, reporting to John Gummer as Secretary of State. Robert Jones, who had a background in the industry, having himself been involved before election to Parliament in the work of the FCEC and National House-Building Council (NHBC), was invited by the Secretary of State in July 1994, at the time of the publication of Sir Michael's review, to indicate what action he would like to concentrate on in the early days of his ministerial career. He took the opportunity to adopt *Constructing the Team* as an important aspect of policy, and encouraged the industry to organise a national conference to discuss and decide on the way ahead.

In a recent interview, Robert Jones made clear that within Government there was unanimity at the highest levels that priority should be given to encouraging the industry to adopt and implement the recommendations in *Constructing the Team*. The one area in which there was inter-departmental disagreement related to the proposal (recommendation 28) for legislation to deal with dispute resolution and liability. According to Jones, the Treasury, Department of Trade and Industry (DTI), and the Departments concerned with Scottish and Welsh affairs accepted that time should be found if possible in the legislative programme to introduce a bill, but this was opposed by Michael Heseltine, who had previously had responsibilities as Secretary of State for the Environment, on the grounds that Government should avoid intervening in an industry's business which the industry should control itself. Jones was able to call on support from some other ministers, notably Michael Forsyth (Scotland) and Tessa Knight in the Treasury, and after high-level discussions it was agreed to proceed with legislation when time allowed. It is the view of Sir Michael Latham that Tony Newton MP (now Lord Newton), as Leader of the House of Commons, chaired the Cabinet Committee which resolved these differences. Although this was a gratifying result, it could be argued that this difference of approach to intervention (dirigiste versus open market forces), manifested within Government, resulted in eventual introduction of a much less comprehensive and radical bill than originally envisaged in the Latham context.

One aspect of the bill related to a main concern of the industry, which was to minimise the worst effects of the traditional adversarial approach to contractual disputes. This was made the substance of Part II of the Housing Grants, Construction and Regeneration Bill, which eventually received the Royal Assent in 1996. Before the act became effective there had to be passed a default position to underpin contracts that did not have clauses that required adjudication. Before the passing of the act, adjudication was not a recognised form of dispute resolution in its own right, but perceived wrongly as merely another type of arbitration procedure. Under the act construction contracts must now include a provision for adjudication, with the adjudicator giving a decision within 28 days of referral. The adjudicator's decision is binding until a final agreement is reached, by consent, arbitration or litigation, or alternatively the parties in the dispute may accept the adjudicator's decision as final. The result of this innovation has been a dramatic reduction in the number and cost of litigious actions throughout the industry. Prior to the introduction of mandatory adjudication, the industry spent more on litigation than on training or research and innovation – an astonishing and depressing fact for a major industry in a developed country.

The Government of the day recognised the importance of the Latham proposals; this was in spite of the risk, as pointed out at the time by Tony Jackson on behalf of the material producers, that there might be ambivalence on the part of Government between the DOE, who as sponsoring department for the industry deplored the industry's continuing fragmentation, and those executive departments such as the Ministry of Defence (MOD) and the Property Services Agency (PSA), who as major spenders might achieve some advantage to their own contracts in the short term from the industry's perceived failure to work together.

With this degree of Government interest, the CIC was able to gain widespread support for the concept of 'a Latham conference' to examine the proposals and map out a programme of action. This duly took place in July 1994, one week after the report's publication, and the programme of proceedings shows how the enthusiasm and excitement engendered by publication of the review ensured the active participation of many of the big names in the industry, whose desire for rapid improvement and change was well known.

The conference agreed to establish a Review Implementation Forum to take forward the 53 recommendations in *Constructing the Team* and to consider the case, and proposals, for establishing a permanent organisation to concern itself with the strategic development of the industry and provide machinery within which the affairs of the industry could be considered on a pan-industry collaborative basis.

This decision was welcomed, and to assist in the necessary work the Department of the Environment seconded an official to work with the Review Implementation Forum's temporary director (Malcolm Dodds) and another temporary staff member Huw Taylor, seconded from NatWest. Their efforts paved the way for the issue in January 1995 by John Gummer of a press release, announcing the establishment of an interim Construction Industry Board. This press release provided the foundation for the emergence of a fully comprehensive board, with representation from all sectors of the industry including clients and the relevant Government departments, with administrative and financial resources to be provided jointly by industry and Government. The work of the board is described in Chapter 5.

In the months prior to the act becoming effective there was a colossal amount of lobbying, most extensively by the specialist contractors who alone caused over 100 amendments to be proposed in Parliament. The Construction Liaison Group (mainly run by the SECG) initiated a massive campaign to persuade Parliament to improve the draft bill to meet their concerns, spending well over £1m. on lobbying MPs of all parties. The help of members in both houses was enlisted to put down amendments, and at the committee stage of the bill in the Commons they secured an amendment to the 'pay when paid' provision that sought to outlaw all contractual clauses that would allow forms of conditional payment. A three-line whip had to be arranged to remove this amendment at the third reading and restore the original wording which simply outlawed 'pay when paid'. There were also attempts to bring in client deposited payments (there had been some bad cases of client bankruptcy) and decennial insurance for projects: as a last-minute measure, three people who had not been involved in these two matters (Will McKee, David Adamson, and Anne Minogue) were asked to make an effort to get agreement to those going into the Bill. On a cold winter night, with heavy rain battering the windows of a claustrophobic committee room, their efforts, which many thought would gain success, were thwarted when one of the clients torpedoed the deal brokered to secure client deposited payments, and then the main contractors, to general surprise, vetoed decennial insurance: sectorism was fighting back.

4 The establishment of the Construction Clients' Forum

Composition; chairmanship; terms of reference; client sectors involved; central/local Government, property, public infrastructure, universities, public/private sector housing; large private sector clients. Strengths and weaknesses; adequacy of resources.

Crucial to the successful initiation of the reforms recommended in the Latham review was acceptance of his personal conviction that the clients should lead, and be recognised as leading, the movement towards integration of the construction process, and its transformation into a customer-focused supply mechanism. 'Implementation begins with clients. Clients are at the core of the process and their needs must be met by the industry' (*Constructing the Team*, paragraph 1.11). This reflected the march of consumerism: just as car purchasers eventually demanded and got cars that would not rust away, predominantly by forsaking British products in favour of foreign, so buyers of buildings were beginning to demand sound buildings at a fair and predictable price.

The trouble was that up to the time of the publication of the review no comprehensive client representative organisation existed. There was no recognised method whereby clients could speak collectively to the supply side of the industry or provide a single point of client contact which the Government could use to pass on its message or seek client participation in Government and industry initiatives. Sir Michael's simple, but strongly worded recommendation:

A Construction Clients' Forum should be created to represent private clients

(Recommendation 1.3)

provided the opportunity for the conference that was set up to consider how to advance the Latham proposals, to press for early establishment of such a body. This received widespread support. Although Sir Michael had envisaged a group which represented private sector clients, it is evidence of how enthusiastically the report was initially received that the central Government part of the public sector of the economy, through the Central Unit on Procurement (CUP) of the Treasury, pressed from the start for public sector clients to be included. Sir Michael has himself said that he did not originally include the public sector in his recommendation because he found it difficult to believe at the time that central Government would accept such a proposal. That they did so is a reflection of Government's early acceptance of his other seminal recommendation (recommendation 1.2) that Government should commit itself to being a best practice client. The way was prepared for the later concept of 'clientship'. The CUP saw the emergence of an effective and comprehensive construction client movement as potentially helpful in its campaign to improve the public sector's performance in commercial matters. The CUP had been established as one of a number of advisory units within the central Government machine. Its objective was to make public servants aware that good practice and performance in commercial areas where public sector bodies were major, and sometimes the largest customers was equal in importance, and in perception of career success, to the provision of policy advice leading to legislation. From being a relatively low-grade activity in the public service, procurement, thanks to the CUP, rapidly came to be perceived as a high-profile executive function, crucial to the success of the Treasury in leading the public sector into the harsher economic world of the 1990s, and a proper place for able civil servants. With this remit, and with Treasury backing, the CUP attracted on secondment experts from the private sector (such as Geoffrey Wort on secondment from Laing), able to communicate their experience and skills in procurement to public servants, whose traditional methods of recruitment and training had up to that time somewhat downplayed its importance.

The CUP allocated one of its full-time officials, Mike Burt, to the newly formed Executive Committee of the Construction Clients' Forum (CCF), and also made available the well-focused advice of Geoffrey Wort to the Forum. This was particularly helpful in showing the clear intentions that the client movement should aim to be practical rather than theoretical in its approach. With this evidence of commitment, the supply side of the industry, through the various umbrella and trade association representative bodies, recognised that the clients were serious in their wish to make progress, and that the supply side's interests would be well served by collaborating with them. The supply side had, certainly in public, welcomed the emergence of a client representative body, both in the conference that followed the launch of the report, and in the Review Implementation Forum that resulted from it. The CIC, with its preponderance of professional bodies, was particularly keen to forge close links with the new Forum, since those of its members who acted in consultancy roles regarded themselves as the clients' representative, whose objectives and interests were synonymous, although many clients welcomed their greater independence and proactivity that flowed from the Latham report: as members of 'the project team' they could talk directly with constructors, not just through the design team and quantity surveyors. The specialist sub-contractors and the material and component manufacturers also saw the Forum as providing a platform for advancing their claims, identified and validated in the report, for earlier and closer involvement with the client in the definition of the clients' requirements and in the execution of the construction process.

The enthusiasm of the public sector clients was, however, only partially matched by that of the private sector. Initially, the Confederation of British Industry (CBI) indicated considerable interest, recognising that their member firms depended on capital investment projects successfully delivered to undertake their production. The CBI nominated the former chairman of their London region, and a CBI council member, Ian Reeves, to contribute to the early work involved in establishing the recommended client body. With the aim of involving the important process of stimulating client action, a loosely formed grouping of off-shore oil and gas operators, the Capital Projects Clients' Group (CPCG), indicated a wish to be involved, and the Construction Round Table (CRT), an informal discussion group of large public and private sector clients, including McDonald's Restaurants and Marks and Spencer, was represented by that body's director.

Largely because they had been so closely involved with the Government in the past, and particularly in the early days when a review of the industry was initially being considered, the property development sector, through their trade association the British Property Federation (BPF), took a major role in establishing the Construction Clients' Forum. The BPF's then assistant director, now director of the British Council for Offices, Richard Kauntze, has suggested (interview 15 July 2003) that BPF members saw the review, and the subsequent proposals for its implementation, as a major opportunity to tackle at long last its outstanding problems in dealing with the industry, particularly contract documentation, liability and risk, and certainty about price and delivery. He, and the BPF's director, Will McKee, felt that the property developer should be at the leading edge of the emergent clients movement, not least because BPF members including construction and development specialists could bring with them detailed knowledge, and widespread recognition among the industry's suppliers that these experienced clients had the authority of their companies to act and deliver.

There can be no doubting the commitment at that time of the BPF as a trade association to the concept of a strong client movement. What is less certain, however, is whether or not the BPF's own member firms were themselves as convinced as their trade association that their long-term interests were best served by allying themselves with the other client sectors, both public and private. Hammerson's, a leading member firm of the BPF, established a public position as a Latham champion, but other leading property developers were considerably more reluctant to show public support for the concept through membership of the new group. There was a view that even leading members of the BPF did not in their day-to-day work demonstrate the principles of the Latham report.

Be that as it may, the BPF offered premises in their London offices for the new client organisation, and seconded the services of their director of policy, Richard Kauntze, to work with the CIPS' nominated officer, Tony Pollington, to initiate the necessary action to bring the client body into existence.

In the early days, however, the general enthusiasm for speedy progress in implementing the Latham recommendations had engendered an 'emphasis' among the organisations which had expressed interest in forming the client group. They had readily agreed to contribute to the establishment of an informal forum, which would be responsible for consulting between themselves, with the aim of producing a recognised client view on issues facing the industry, and which were to be considered collectively within the Construction Industry Board.

The Construction Clients' Forum (CCF) was officially launched at the end of 1994. The originating member bodies were:

British Property Federation
Capital Projects Clients' Group

Central Unit on Procurement/HM Treasury
Chartered Institute of Purchasing and Supply
Committee of Vice Chancellors and Principals (of Universities)
Confederation of British Industry
Construction Round Table
Highways Agency
Ministry of Defence (Works)
National Housing Federation (formerly National Federation of
  Housing Associations)
National Health Service (Estates)
Department of the Environment (observer)

The first chairman, elected by the initial members, was Stuart Humby, director of purchasing at NatWest, and a former president of the Chartered Institute of Purchasing and Supply, which had been closely involved in the deliberations with Sir Michael Latham about the involvement of the clients in his review.

By the end of 1996, the Local Government Association had agreed to participate, and by the end of the first year an executive committee was directing a small office in establishing the client movement as capable of contributing a representative client voice to the industry's collective discussions. This initial executive committee was expanded the following year (1996) to reflect the growing representation:

Chairman: Geoffrey Wright (Director, UK Property, Hammersons
  plc)
Deputy Chairman: Ian Reeves (Chairman, Carolian International
  Consultants Ltd and Former Chairman, CBI London Region)
Deputy Chairman: Terry Rochester (Chief Highway Engineer, High-
  ways Agency)
Past Chairman: Stuart Humby (Director of Group Purchasing,
  NatWest Bank)
David Adamson (Committee of Vice Chancellors and Principals;
  Bursar, University of Bristol)
Henri Pageot (Executive Director, Construction Round Table)
Robert Soar (Head of Contracts, Property Department, Hertford-
  shire County Council)
Will McKee (Director General, British Property Federation)
Executive secretaries: Anthony Pollington (Head of Public Affairs,
  Chartered Institute of Purchasing and Supply)
Richard Kauntze (Assistant Director, British Property Federation)

The newly formed CCF adopted a formal statement of objectives, which was widely publicised.

Aims and objectives

The CCF Mission

To represent the interests of construction industry clients collectively by:

- encouraging clients to achieve value for money through best practice;
- securing major and measurable improvement in the performance of the supply side of the industry, including cost reduction;
- promoting policies which can achieve a healthy, stable and skilled industry which is competitive, well capitalised and competent.

Our objectives

In fulfilling this mission we will . . .

In the short term

- Identify indicators to measure performance and establish benchmarks, by ensuring high quality client input to the work of the CIB panel on Cost and Productivity;
- assist small and occasional users of the services of the industry to recognise and use their market strength to achieve improved performance;
- promote the use of the CIB and other good construction codes of practice;
- establish a mechanism to allow clients to share information;
- raise clients' awareness of the benefits of partnering as a significant option for procurement;
- contribute to and monitor the development of construction legislation and regulation which affect clients' interests in the industry, vigorously resisting measures detrimental to clients' interests;
- ensure that clients effectively influence new forms of contract;
- ensure that clients' interests are adequately represented in ministerial briefings;
- strengthen client involvement in the strategic direction of construction research;

- support proposals by CCF members for research projects which will benefit clients generally;
- ensure a strong and co-ordinated client input into new initiatives.

In the medium to long term

- Take steps, if necessary alone, to simplify the consultative structure of the industry;
- contribute to the development of a register of client advisers and contractors;
- instigate research to promote efficiency, including evaluation of partnering;
- establish mechanism to enable feedback to industry on performance, perhaps utilising *Which*-style reports;
- benchmark performance against an ultimate goal of 'zero defects'.

In all our work our approach will be to ...

- Work in collaboration with the supply side of the industry whenever we can but be prepared to 'go it alone' if necessary;
- constantly strive to eliminate poor practice;
- place emphasis on using the skills and strengths of CCF member bodies to the full and only launch central initiatives if there is no way of proceeding using existing structures;
- focus on issues of interest to all clients but also encourage action/initiatives by specific client groups;
- avoid being drawn into participation in initiatives which are not central to client needs.

What we can offer

CCF recognises that membership should give added value. Our aim is to provide:

In the short term

- An advice line linking to advice centres;
- examples of value for money obtained through new approaches;
- access to our existing communication network (CIB, Government, contractors, etc.).

In the long term

- A *Which* approach to review of the industry;
- a suite of client advice leaflets;
- reference costs for types of construction (examples of Latham-type projects making savings);
- ability to influence research and development;
- endorsement of training programmes, qualifications or client advisers;
- 'museum' of bad examples;
- news-sheet.

The movement's relative success or failure in meeting these aims and objectives will be examined later (Chapter 7), but how truly representative of the total client movement was this early grouping?

In terms of the client sectors represented in the Forum, the member organisations brought within their ambit nearly 80 per cent of identified client demand (in terms of quantum of money spent) on the industry (excluding DIY) in 1996, as shown in Figure 4.1.

On the face of it, this gave the movement considerable credibility in claiming to speak authoritatively for clients in relation to the supply side of the industry and to Government. In reality, however, this is somewhat misleading, since the member organisations were themselves 'umbrella' representative bodies (BPF, CIPS, CBI, Local Government Association (LGA)) and it proved much more difficult to obtain enough specific input of the client organisations actually

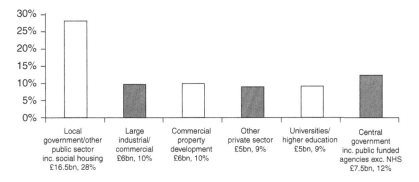

Figure 4.1 Percentage value of total annual client demand (1996) by client sector represented on the CCF. The total value of client demand is approximately £58bn, and the value of demand by CCF members is £46bn.

engaged in construction contracts, and responsible for paying the bills. Only three members of the executive were themselves responsible for signing construction contracts. This initial weakness in the CCF's structure, although perhaps unavoidable in reality, proved difficult to overcome, and undoubtedly debilitated the movement in later years, as will be shown subsequently.

This group was able to call on client representation from a wide selection of public and private sector client organisations. As noted earlier, Stuart Humby, director of procurement at NatWest, had agreed to act as the CCF's first chairman, and at an early meeting at the bank's offices in Lothbury well over 30 representatives of client organisations attended. As the agenda for that meeting shows, the CCF was sufficiently well established by January 1996 to be confident in formulating programmes of work aimed at contributing to the pan-industry work within the CIB concerned with following up the main issues identified in *Constructing the Team*. Undoubtedly the impetus behind this was the involvement of influential private sector 'big hitters' such as Northumbrian Water, Sainsbury's, British Telecom, Prudential, National Power, BAA, Hammersons, Slough Estates, and Marks and Spencer, as well as major public sector client organisations. Through the involvement of these organisations, the CCF was able to call on many individual experts and nominate them to participate on behalf of clients in the specific work programmes initiated by the CIB.

At this meeting in January 1996, the CCF also adopted a constitution, drafted for it by the officer allocated to its secretariat by the BPF, Richard Kauntze. This constitution emphasised the informal nature of the Forum, and its determination to avoid, if possible, a rigid and bureaucratic structure. Stuart Humby was succeeded as chairman in April 1996 by Geoff Wright, director of construction of Hammerson's plc. Up to that time the 'fledgling' CCF had been supported entirely by the two organisations, the BPF and the CIPS, who supplied administrative staff help, and, in the case of the BPF, office premises and office support. Both these organisations, having calculated the value of their input to that date as of the order of £70k indicated that they were unwilling to continue to accept this open-ended commitment. Stuart Humby and his successor, Geoff Wright, therefore sought and obtained the agreement of the CCF members to contribute up to £5k each annually, providing an income estimated to total some £45k for the year 1996. This was, from the outset, recognised as being well below the full costs involved in running the CCF, but the general goodwill, enthusiasm, and optimism of those early members suggested that confidence in achieving considerably higher contributions would not be misplaced. Already, large numbers of busy people were giving several hours per month of their time, free, to advance the reform: this reflected the rapid growth across the industry in personal commitment of high-level industry leaders.

The early level of enthusiasm was not in fact retained as the years passed. The reasons are difficult to define, even with hindsight, but seem to relate in some degree to the inability, or unwillingness, of the important private sector clients to ensure the continuity of their individual representatives in the CCF's ongoing programme, and in the planning of its future organisational development. There was a widely held view among top management that as construction was not part of the organisation's core business, it could be delegated to specialists.

Despite the almost universal acknowledgement of the importance of the review's recommendation that clients should come together in a truly representative forum allowing them to speak with authority, both to Government and to the supply side of the industry, the Forum proved only partially able to bring this about. Although there had been some initial positive indications from private sector clients of their willingness to join the CCF, particularly in the retail industries, who were members of the Construction Round Table (CRT), the CRT itself held aloof and that was damaging. This may have been partly because of suspicion on the part of Forum members that the CRT's membership retained close contacts within the CRT's programme of work with particular contractors, and that this might prejudice the Forum's claim to be a purely client-based group. What seems to have been more influential, however, was the feeling among some CRT private client members that their ability to consult each other on commercial matters connected with their construction contracts, in the informal and confidential atmosphere of the CRT, would be in some way compromised if their representatives were subject to the inevitably more structured proceedings of the Forum and the Construction Industry Board.

The result was that attendance both at formal Forum meetings, and at executive committee meetings, where the future direction of the client group was formulated, declined quite seriously during the first half of 1996. The CBI's representative, Ian Reeves, pressed hard for full involvement of CRT members, since he saw that evidence that major players among private sector clients were contributing to the client movement would encourage the CBI and its members to participate. He was, however, unable to get the full backing of the CRT's director at the time, Henri Pageot, and hence the co-operation of its member firms. David Adamson, who represented the Committee of Vice-Chancellors and Principals of UK Universities, and who was in charge of a large construction programme at Bristol University, continued to devote considerable time to the Forum's affairs, both as one of the successor deputy chairmen of the Forum, and in collaboration with the officers of the British Property Federation, where the Forum was housed. Nonetheless, attendance at executive committee meetings became so attenuated, sometimes involving only those mentioned above, the chairman and the Forum's officials, that the Department of the Environment's officers who kept a watching brief on the development of the client movement, to which they gave considerable importance in their plans for the future of the industry post-Latham, took to attending meetings of the CRT in preference to those of the CCF. At one period the future looked so doubtful that David Adamson seriously considered taking the Committee of Vice Chancellors and Principals (CVCP) out of the CCF, a move which would have at that time brought about its demise. That he decided not to do so, and that the movement in fact developed a renewed strength in the latter part of 1996 and the early part of 1997, resulted from its decision to devote most of its intellectual resources to defining and publicising precisely what clients wanted from the industry and how this might best be provided. The emergence, and subsequent influence, of this 'Client's Manifesto' is described in Chapter 7.

5 The establishment of the Construction Industry Board

Ministerial presidency; support of Gummer/Jones; chairmanship of Sir Michael Latham; respective position and involvement of industry sector interests; employers, sub-contractors, professionals, material and component manufacturers; allocation of seats on the board; background to appointment of Chairman (Ian Dixon) and Director (Don Ward); early policy papers, position statement, proposed work programme. Embodiment in DOE.

The post-Latham review conference (1994), and the Review Implementation Forum which resulted from it, forced the industry and Government to consider whether some more permanent strategic body for the industry was needed. As noted earlier, the Review Implementation Forum's task was defined, following the conference, as twofold:

- to take forward the 53 recommendations in *Constructing the Team*; and,
- to consider the case and proposals for any permanent pan-industry strategic organisation.

The declared support of Government encouraged the industry to embrace the idea of a permanent body, enabling the then Secretary of State John Gummer to issue his press release in January 1995. Establishing an interim Construction Industry Board. The Secretary of State agreed to accept the designation of President of the Board. Malcolm Dodds, at that time on the staff of DOE, was seconded as director.

The decision to set up a formal board was not initially universally popular, although given the momentum behind the proposal from Government, no sector of the industry wished to be publicly identified as opposing it. The personal commitment and determination to pursue the Latham recommendations of the senior civil servant concerned, Phillip Ward, head of DOE's Construction Directorate, has already been remarked on, and is a good example of how even the most powerful industry can be persuaded to adopt a course of action if the particular official possesses the necessary skills and strength of leadership. This is not to suggest that his approach was immune from criticism. Interviewed years later, representatives both of the contractors and of the Board's officials expressed the view that the early activity concerned with identifying the need for, and possible structure of, a pan-industry body was heavily civil service dominated, and that this culture influenced the subsequent activities of the Board, not always to its advantage. However, there was not at the time, nor subsequently, any criticism that Ward's approach reflected any political objective on behalf of the Government, nor that it was dictated by other than a conviction that early action was imperative.

Possible alternatives to the concept of a board were considered. Phillip Ward has described detailed discussions, both within and outside the Department, with Sir Michael Latham about the advantages and disadvantages of such a structure. Indeed, reflecting the view at the time of some industrialists that effective action was more likely to result from having a strong, authoritative individual, capable of imposing a programme of action on the industry, rather than from a board with its connotations of bureaucracy, serious consideration was given to the appointment of such a 'construction industry czar'. Phillip Ward (interview 9 July 2003) has made clear that, irrespective of the difficulty of finding such a person and of their acceptability to all parts of the industry, history has shown that the effectiveness of such appointees in other sectors of the national life cannot be guaranteed.

In their evidence to the Latham review, the CIPS had proposed a separate organisation, called Construction Sourcing, which would have acted in a commercial environment by amalgamating aspects of demand, in order to smooth out peaks and troughs of supply and demand, and allow the clients and their supplier to plan for stability. This was not proceeded with, largely because of doubt on the part of the supply side of the CIPS' ability to deliver such a concept. The effect of such an innovation on the industry, had it been successfully introduced, can only be a matter of conjecture.

Be that as it may, the interim board moved swiftly in appointing

Ian Dixon CBE, at the time chairman of Willmott Dixon and also chairman, in view of his civil engineering interests, of the Construction Industry Council (CIC), to develop proposals and a formal business plan for a permanent Construction Industry Board. His revised business plan was presented to the Board in June 1996, after the six member bodies involved in the interim board had formally agreed to participate.

The six member bodies were: the Construction Clients' Forum (CCF, representing regular and occasional/one-off clients), the Construction Industry Council (CIC, representing professional institutions, specialist standards-setting bodies and the professional services), the Construction Industry Employers' Council (CIEC, representing main contractors), the Construction Liaison Group (CLG, representing sub-contractors and specialist trade contractors), the Association of Construction Product Suppliers (ACPS, representing the materials and product suppliers), and Government (including DTI, Scottish Office and the Health and Safety Executive, with DOE as lead department).

Sir Ian Dixon's (he was knighted in the 1996 New Years' Honours) definition of CIB's mission was:

> to provide strategic leadership and guidance for the development and active promotion of the UK construction industry, through liaison between representatives of the construction industry, its clients, and Government, in order to improve efficiency and effectiveness throughout the construction procurement process. CIB aims to secure a culture of co-operation, teamwork and continuous improvement in the industry's performance. As implied above, its principal objectives are to implement, maintain, monitor and review the recommendations of *Constructing the Team*.

He defined the CIB's aims and objectives as follows:

> CIB aims to secure a culture of co-operation, teamwork and continuous improvement in the industry's performance.
>
> CIB's principal objectives are to implement, maintain, monitor and review the recommendations of Sir Michael's report, *Constructing the Team*, in particular to:
>
> i deliver improved construction performance, measured in terms of its quality, productivity and competitiveness;

ii   reduce conflict and increase trust;

iii   improve security of payment throughout the construction process; and

iv   secure a 30% cost reduction in real terms through improved productivity and procurement methods by the year 2000.

The Board acts as a forum at a strategic level, co-ordinating and facilitating the activities of its member bodies and others who operate as deliverers. It seeks to move forward on the basis of consensus wherever possible, without compromising its ability to address potentially difficult or contentious matters. It seeks to add value to existing industry mechanisms and structures, and will not duplicate or add a layer of bureaucracy.

He identified the Board's core activities as:

- measures designed to improve competitiveness and productivity;
- research and innovation;
- practice procedures and codes of practice (including implementation);
- human resources of the industry;
- relations with Government and within the industry;
- external image of the industry;
- legislative and regulatory matters.

Membership of the Construction Industry Board initially caused some problems. The main contractors, represented by the Construction Industry Employers' Council (CIEC) pressed for a greater number of seats on the Board than other supply side parties, on the grounds that traditionally the main contractors were the recognised supply side parties in construction contracts. On the other hand, as pointed out by Tony Jackson, chief executive officer of Blue Circle Cement Properties, and a leading figure in the Association of Construction Product Suppliers (ACPS) the umbrella organisation representing material and component manufacturers and suppliers, in terms of the value of annual turnover, Blue Circle alone exceeded the total annual turnover of the ten largest main contractors together.

What was agreed by all parties, however, was that, in recognition of the acceptance across the industry of the Latham conclusion that the industry should be 'client-led', the clients' representation on the Board should reflect this position. Accordingly, Sir Ian Dixon recommended, and the Board accepted, the following structure and allocation of seats:

The Construction Industry Board (CIB) will comprise representative umbrella bodies of the construction industry, its clients (both private and public sector), and Government.
Initially the member bodies of the CIB will be:

- the Construction Industry Council (CIC);
- the Construction Industry Employers' Council (CIEC);
- the Construction Liaison Group (CLG);
- the Construction Clients' Forum (CCF);
- the Association of Construction Product Suppliers (ACPS); and
- the Government led by the Department of the Environment (DOE).

This will consist of four representatives each from the umbrella bodies (except the clients, who will have five as they also represent all aspects of the public sector). The Board will normally meet four times a year to set policy and identify priorities. Its specific tasks will include agreeing and reviewing the business plan, and receiving reports on the state of the industry and other matters.

The Secretary of State for the Environment will be CIB's president. The Board will select a chairman, whose role is to chair the Board and to represent the CIB to all its audiences, and two deputy chairmen.

Hence representation on the Board was agreed as shown in Table 5.1

Sir Michael Latham agreed to be the CIB's first chairman, on a part-time remunerated basis (nominally two days per week),

Table 5.1 Industry representation on the CIB

CIB member body	Number of seats
Chairman and deputy chairmen	3
CIC, CIEC, CLG, ACPS	4 × 4
CCF (including all aspects of the public sector)	5
Government (as sponsor and regulator)	4
Total	28

supported by a small central secretariat to be financed from subscription of the member bodies, and initial supporting grants in aid from the industry's sponsoring department, DOE.

The effort put in by Sir Michael Latham, with the support of DOE, in publicising the results of, and seeking support for, his review and its recommendations was truly astonishing. Between the end of July 1994, when *Constructing the Team* was published, and June 1996 when he handed over the CIB chairmanship to Sir Ian Dixon he attended, and spoke at, no fewer than 33 conferences held by companies and organisations in their own premises, and 60 public conferences held by particular industry sectors. He managed to eat his way through over 50 official lunches and dinners, at which he was often the chief official guest, reflecting the level of interest his report had stimulated. Echoing his view that the reforms the report proposed are as important to the small and occasional client as to the major firms in the industry, he willingly accepted invitations to speak to organisations as diverse as associations of construction lawyers, management consultants, technical magazine editors, trade associations of specialist suppliers, professional institutes and housing associations.

In due course, Sir Ian Dixon proposed a budget as summarised in Table 5.3.

As regards the CIB's income, this was envisaged to come primarily from subscription from the member bodies (see Table 5.2). Sir Ian envisaged this basic income being augmented by revenue from publication sales, conferences and other initiatives in the longer term.

With the acceptance by the main players of this basic structure, the situation allowed the CIB to recruit a chief executive. The post, publicly advertised at a circa £60k package (1996 salary levels), attracted wide interest, with a shortlist of six emerging from over 70 applicants. The chief executive's job description is given on pages 46–47.

Table 5.2 Proposed 1996 income

CIB member bodies	Annual subscription 1996 (£k)
CIC, CIEC, CLG, ACPS @ £20k each	80
CCF	20
Government (DOE)	120
Other income (publication sales etc.)	5
Total	225

Table 5.3 Sir Ian Dixon's proposed budget for the CIB

Item	1996 budget (£k)	Notes on 1996 budget	Likely out-turn 1995 (£k)	Notes on 1995 out-turn 1995
Chairman	12	Assume 1 day/week		Sir Michael included in next item
Chief executive and other staff	160	(total package incl. benefits and NI contributions) Chief exec+1 senior, 1 junior and 1 secretary, all full-time	198	Currently includes Sir Michael Latham and 3.4 secondees comprising 4 civil servants+1 NatWest employee part-time; shared with Construction Procurement Group
Office rental	20	Assume £20/ft^2 fully inclusive	5	No charge at Building Centre, July–Dec
Phone/fax			4	
Office expenses	33	No provision for contingencies	7.5	Stationery, postage, photocopying, meetings
Travel expenses			3	
Equipment, etc.			2	Mainly IT
Total	225		220	In 1995, DOE contributed £136,754, three supply side bodies gave £20,000 each, CCF gave £22,205

CONSTRUCTION INDUSTRY BOARD

RECRUITMENT OF A CHIEF EXECUTIVE

Job description

Potentially a position of outstanding importance and high profile in the UK construction industry. As chief executive of the Construction Industry Board this person would be a key player in the work of this pan-industry forum. The mission of the Board is to provide strategic leadership and guidance for the development and active promotion of the UK construction industry through liaison between representatives of the construction industry, its clients, and Government. A negotiator and listener at heart, the applicant will have proven ability to communicate persuasively to all levels of audience. The capability to lead a team of four in a 'hands-on' way on day to day issues and long term objectives is a prerequisite, and an understanding of the main concerns of the industry and its clients is desirable, together with the issues addressed in Sir Michael Latham's final report *Constructing the Team*. This is a fixed-term contract (renewable).

Key responsibilities

The following aspects of the proposed position will be considered when selecting applicants:

A. Hands-on leader of a team of four.
B. Developing a new organisation.
C. Increase public awareness and image of the construction industry.
D. Key role in the future of the construction industry.
E. Responsible and accountable to the board and a member of the executive committee.
F. Build relationships between member bodies and industry.
G. Dealing with day to day issues and the long term development and prosperity of the industry and the organisation.
H. Lead the marketing of the board.
I. Securing financial support to implement initiatives.
J. Attend board meetings *ex-officio*.
K. Meeting set objectives made in the business plan.
L. Act as company secretary of the board's service company.

Personal attributes

The following personal attributes are desirable:

A. A hands-on person, able to work in a small team.
B. A good listener and negotiator.
C. Decisive and determined to make things happen.
D. Remaining sympathetic and pleasant whilst under pressure to perform.
E. Sound knowledge of the construction industry and relevant issues.
F. Excellent communicator to all levels of audience.

The decision on appointment was made ultimately by the CIB, on the recommendation of an interviewing panel with strong representation from DOE. Their choice fell on Don Ward, the acting chief executive, who was originally a principal scientific officer in the civil service, and had transferred to the Administrative Branch under the Scientific and Professional Administrative Training Schemes (SPATS), designed to attract scientific and professional grade officers in the civil service into the Administrative Branch. Having been acting chief executive, Don Ward formally took up office in the Board's premises, rented from the Building Centre, in July 1996.

The relative success of the Board in the form in which it eventually emerged is discussed later (Chapter 6 and following). With hindsight, Phillip Ward (interview 9 July 2003) certainly felt that its composition, made up as it was predominantly from trade association representative bodies, was not wholly representative of the interests involved. He notes, however, that it would have been unrealistic to look for membership of individual chief officers from commercial companies in the industry. Such people have neither the time nor the motivation (nor in reality the specific technical expertise) to involve themselves in the day-to-day technical issues of the industry, with which the Board would inevitably find itself concerned. Again with hindsight, he now feels that the Department could have been more demanding on the constituent parts of the industry to start action in support of the Latham recommendations, which they had all bought into, rather than wait for nearly two years for a formal consensus on the way ahead to emerge as the Construction Industry Board. That said, it is important to recognise the huge amount of effort and useful work in developing cross-industry relationships, and the documents later to be issued as well-discussed, reputable and influential codes of good practice. For the first time, the industry and its clients had been galvanised into wide debate and improvement in the way many of its people actually did their business. Already, very senior and busy members across this huge, diverse and fragmented industry had freely and enthusiastically given hundreds of thousands of hours of their time to improving their common lot: the giant that had twitched, but never moved purposefully, had started to walk.

6   The CIB in action

Early meetings and agenda; approach to the Construction Industry Bill, the CIB Working Groups 1–10; publication of the codes
of practice (the 'Six Pack'); Sir Ian Dixon's chairmanship; chairmanship of Tony Jackson; chairmanship of Chris Vickers; relationship with other bodies, e.g. Design Build Foundation (DBF),
Reading Construction Forum (RCF), European Construction
Institute (ECI), etc.; relationship with Government (independence of, etc.); mechanism for control of 'self-employed'; concept
of quality/price balance; construction line; Working Group
11–13; partnering; project insurance; 'The Scheme' (for adjudication); what linkages were there with the realpolitik of business;
what ability to understand and respond to blockers and enablers
to meet the objectives described in Chapter 1?

The Construction Industry Board (CIB) was officially launched in
February 1995. Its membership, as noted earlier, was:

Officers
President: The Secretary of State for the Environment
Chairman: Sir Michael Latham; Sir Ian Dixon (from 24 July 1996)
Deputy Chairmen: Sir Ian Dixon (succeeded after July 1996 by
Tony Merricks); Roger Squire

Member bodies
Construction Clients' Forum (CCF)
Construction Industry Council (CIC)
Construction Industry Employers' Council (CIEC)

Construction Liaison Group (CLG)
Association of Construction Product Suppliers (ACPS)
Government departments (DOE, Treasury/CUP, Scottish Office, Welsh Office).

Immediately after publication of *Constructing the Team*, the Review Implementation Forum set up a number of pan-industry working groups to implement the Latham recommendations. These recommendations fell into four main categories:

1   Recommendations for clients (public and private sector):

   - publish a Construction Procurement Strategy Code of Practice;
   - publish a guide to client briefing;
   - promote a mechanism for selecting consultants on quality as well as price;
   - establish a client forum;
   - establish Government as a best practice client.

2   Recommendations for the industry:

   - adopt a target of 30 per cent real cost reduction by year 2000;
   - improve tendering arrangements/registration (with Government);
   - draw up a joint code of practice for selecting sub-contractors;
   - implement the recent reports on training and on the education of professionals;
   - improve public image;
   - produce co-ordinated Equal Opportunities Action Plan.

3   Recommendations relating to contracts:

   - develop standard contract documentation based on a set of principles (including independent adjudication, pre-pricing of variations and Trust Accounts for payments);
   - produce a complete standard family of interlocking contract documentation;
   - contract committees – restructuring;
   - recommendation for increased use of New Engineering Contract.

4   Recommendations for legislation:

- introduce legislation against unfair contracts;
- introduce legislation to underpin adjudication and trust account proposals;
- implement DOE working party proposals on liability legislation;
- introduce mandatory latent defects insurance.

In relation to category one (clients), the Client Forum was duly set up, and the Government took a number of policy initiatives and internal examinations of public sector practice to advance its claim to become a best practice client. These included:

- a review of construction procurement by Government under-taken as an efficiency scrutiny report by the Cabinet Office unit headed by Sir Peter Levene (later Lord Levene), the Prime Minister's Advisor on efficiency;
- Ministry of Defence: management of the Capital Works Pro-gramme (a National Audit Office report on the MOD's capital works programme);
- 'Setting New Standards: A Strategy for Government Procure-ment' (a Government White Paper on procurement);
- 'Construction Quality: A Strategy for Quality in Construction' (a consultation document issued by the joint Government/industry Quality Liaison Group).

The recommendations for client guidance on procurement, briefing and selection of consultants were taken forward by a number of working groups of the Board, as were the recommendations for action by the industry (Category 2). Recommendations for action on contract documentation (Category 3) were referred to the standing bodies of the industry, the Joint Contracts Tribunal (JCT) and the Construction Contracts Standing Joint Committee (CCSJC), which were asked to pursue the concept of a family of standard contracts for the industry and the possibilities for rationalising the structure of the contract writing bodies. The Institution of Civil Engineers speed-ily issued a second edition of the New Engineering Contract (the New Engineering and Construction Contract), drafted to incorporate the main principles of construction contracts which had been set out in *Constructing the Team*, and which were intended to promote a col-laborative rather than adversarial approach to contracting.

The pattern of working groups adopted by the CIB is expressed graphically in Figure 6.1. The groups were genuinely representative of all sectors of the industry in their membership. They were given objectives to achieve, under the designated leadership of individuals as follows:

> *WG1: Briefing the team*   (Chairman: Frank Duffy (CIC))
> Define the process; what affects the quality of briefing; how can good briefing be benchmarked?
>
> *WG2: Code of practice for clients*   (Chairman: Phillip Ward (DOE))
> Assist clients to obtain value for money; define best client practice.
>
> *WG3: Sub-contractor selection*   (Chairmen: Chris Sneath (CLG) and Chris Vickers (CIC))
> Selection on price and quality; develop team-working; reduce tender lists.
> (WG3a: Code of practice for selection of main contractors)
> (WG3b: Code of practice for selection of sub-contractors)

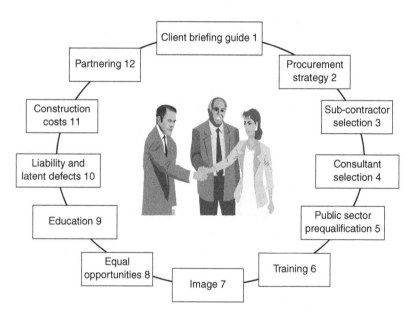

Figure 6.1 The CIB working groups, 1994–1996.

WG4: Consultant selection (Chairman: Geoff Wright (CCF))
Define quality/price mechanism; investigate single consultant register for public use.

WG5: Public sector pre-qualification (Chairman: Geoff Wright (CCF) and Ted Evans (CIEC))
Develop standard pre-qualification form for public sector work and management information system.

WG6: Training the team (Chairman: Hugh Try (CIEC))
Define precisely what initiatives are needed from the public sector and industry.

WG7: Image of the industry (Chairman: Martin Laing (CIEC))
Improve industry's image; set target for measuring progress.

WG8: Equal opportunities (Chairman: Ian Dixon, later Sandi Rhys-Jones (CLG))
Industry's image puts off good recruits; need to attract women to the industry.

WG9: Training construction professionals (Chairman: Michael Romans (CIC))
Needs of post-chartered professionals and possibility of rationalising professional qualifications.

WG10: Liability and latent defects insurance (Chairman: Roger Squire (CCF))
Pursue recommendations on compulsory Latent Defects Insurance.

WG11: 30 per cent reduction in construction costs (Chairman: Peter Alden (CCF))
Demonstrate how savings can be achieved and how they can be measured.

WG12: Partnering (Chairman: Charles Johnston (CCF))
Promote best practice; establish partnering benchmarks; develop appropriate training and education.

As regards proposals for legislation (Category 4), most clients represented in the CCF totally rejected the idea of mandatory latent defects insurance and the supply side, principally the main contractors, would not go along with the introduction of accountable trust funds as a guarantee to clients of successful performance. The Government's proposal on legislation making defect liability insurance mandatory was referred to the Law Commission for further consideration. The result was to lead to a somewhat emasculated Government bill on construction legislation, which was introduced in Parliament late in 1996 as Part II of a Housing, Construction and Regeneration Bill which, partly as a result of pressures on the Parliamentary timetable, incorporated a number of measures relating to housing and planning. Part II of the bill relating to construction provided for:

- statutory entitlement to independent adjudication;
- statutory entitlement to payment by instalment;
- statutory entitlement for a 'final date for payment' of each instalment to be specified;
- advance notification of set-off;
- no enforcement of 'pay-when-paid' clauses (except in cases of insolvency);
- entitlement to suspend performance of contract where payment is not made by due date or other party fails to comply with adjudicator's decision.

Part III introduced a measure, not proceeded with, to provide for the Statutory Registration of Architects.

Sir Michael Latham expressed the view that, if only one of his recommendations was to be successfully adopted, it should be that on adjudication.

All the working groups reported to the CIB by the end of 1996, and their reports were published individually and as a complete boxed set of CIB publications at an inclusive price of £65. Over 1,000 sets were eventually sold, and they remain definitive statements of the detailed industry issues identified in *Constructing the Team*. More copies of Report 4a on how to balance quality and price were sold than all others together. Although written in simple language and aimed at clients as well as industry professionals, they are, to this day, to be found in many offices of main and sub-contractors, professional advisers, public sector organisations, and consumers' associations and individual clients.

How successful were they?

There is no doubt that in the period from February 1995, when the CIB was launched with Sir Michael Latham in the chair, and in the period following his chairmanship, when Sir Ian Dixon took over as chairman with Roger Squire and Tony Merricks as his deputies in July 1996, the Board successfully pushed forward completion of the work programme defined at the conference which launched the Latham review. The Board's brief was to complete the 12 reports described earlier, and then to implement action programmes to get the industry to change its way of working in order to reap the benefits which these reports, and the pan-industry working groups that prepared them, had shown were within the industry's reach. In order to ensure continuity and preserve momentum the Board consolidated the work of the groups into three main policy sections, establishing the management structure illustrated in Figure 6.2.

It is the view of the CIB's then chief executive, Don Ward (interview 12 June 2003) that of these three 'improvement panels' – productivity and cost improvement, good practice, and registration systems – the most successful outcome in the long run related to the work on improving the image of the industry, which featured as part of all three panels' programmes. This resulted, inter alia, in the introduction in October 1997 of National Construction Week (NCW), launched with the financial and organisational support of BT at the BT Tower, and in the following year at the City of London Boys School, and which obtained considerable press and media coverage, at least within the industry. This interest was stimulated by the

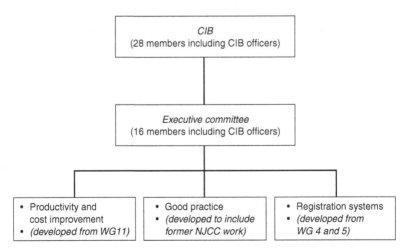

Figure 6.2 Three sub-groups were set up to take forward working group recommendations and other CIB issues.

involvement of Hammerson's, whose construction director, Geoff Wright, sometime chairman of the client's body the CCF and subsequently a deputy chairman of the CIB, agreed to chair the organising committee for NCW. While quite widespread in the industry itself, this level of interest was not at first reflected among the national media or the public at large. In Don Ward's view, the lack of resources available to the Board meant that NCW, and indeed other achievements of the three panels, was seriously undersold in NCW's first couple of years, with the result that many of the discreditable myths about the industry remained embedded in the public mind, whether justifiably or not, for a long time after the industry had committed itself to improvement. NCW did go on to grow in importance. (It is now run by the Construction Industry Training Board (CITB) on behalf of Construction Skills, the sector's skills council.) Another successful innovation was the establishment with major financial support from the Department of the Environment and the Chartered Institute of Building (which also managed the scheme on behalf of the CIB), of the Considerate Constructors Scheme, now adopted widely across private and public sector construction sites throughout the UK via local councils.

The Good Practice Panel brought together large numbers of experts who contributed considerable time and expertise to the preparation and provision of guidance to clients and suppliers. Even at this early stage of the decade of reform, personal hours to the value in the excess of £$\frac{3}{4}$m. per year were being donated to the work of CIB (that figure was to grow massively in the late 1990s). Although Don Ward remains critical of the real effect that this work had in the early days, it led ultimately to the consolidation of the Construction Best Practice Programme (CBPP), and to the encouragement to clients to define and implement programmes for their own improvement (see Chapter 7). It also focused the attention of the industry, including clients, on such concepts as value management, quality/price balance, partnering, and risk management, and if improvements in practice in these areas were not readily apparent in the early days, their importance became accepted in the industry's collective mind, largely as a result of the panel's work.

The Productivity and Cost Improvement Panel (PCIP) was established primarily to put 'meat on the bone' of the industry's target to achieve a 30 per cent cost reduction/productivity improvement. Despite general agreement that improvement was overdue, from the start there was considerable cynicism both about the grounds for defining this target and the realistic possibility of ever achieving it. Its success depended on the adoption by all sectors of good practice, and the ability to define methods of measurement and indicators that would show progress towards reaching the target. The attitude of some parts of the supply side towards the industry's ability ever to obtain the improved performance that was being talked about at the time was summed up by one contractor's remark to Don Ward, 'I'll believe it when I see clients like Hammerson's changing their behaviour.' Nonetheless, the panel's work resulted in the industry accepting that industry-wide indicators of performance were the only way of showing how the commitment to improvement was resulting in a better product. The clients particularly, through the representatives that the CCF provided on this panel, pressed strongly for industry key performance indicators to be designed and implemented and KPIs were eventually developed further, and adopted. (The concept of KPIs were to get a huge boost post-Egan.)

In the spring of 1997 Sir Ian Dixon, who had chaired the board for just over a year, succeeding Sir Michael Latham, announced that he would retire early, due to ill health. This was a serious blow to the CIB at this time, since Sir Ian, with his experience as chairman of a major construction company, and with his knowledge of the civil engineering profession, had proved very acceptable to all sectors and was widely admired. He had also enjoyed good relationships with the Conservative Government, as a result of his own political work in local government and for the party.

Finding a successor did not prove easy. The position had been established in the CIB that while the Board's two deputy chairmen should be nominated respectively by the client movement as represented by the CCF, and by the supply side as represented by the CIEC, CLG, CIC, and Construction Products Association (CPA) in turn, the chairman should be a nationally recognised figure without overt fealty to any particular sector interest. The chairman was also expected to enjoy the support of the Government, advised by the Department of the Environment.

The customary processes of consultation were severely compressed in time; the choice fell on Tony Jackson, previously chief executive of Blue Circle, although he had been retired from that business for some two years. Sir Ian Dixon contacted Tony Jackson at his holiday home in Spain. In a conversation designed to convince him to accept nomination, Sir Ian emphasised to Tony Jackson how it was essential that a chairman should be found who was both recognised as an active commercial figure in the industry and who would be regarded as sufficiently objective to be acceptable to the six member organisations of the Board whose interests were from time to time in conflict with each other.

Tony Jackson accepted the challenge, although he indicated subsequently (interview 28 August 2003) that he entertained some doubts about his acceptability to the client sector. He recognised also that he might be in a relatively weak position in comparison with his predecessor, in that he was unsupported by an organisation and would have to rely entirely on the office and staff of the Board itself. He accepted the position, however, showing the extent of his commitment by cutting short his holiday in order to preside over the Board's May 1997 annual general meeting. At this meeting his concern about the position of the client movement in relation to his acceptance of the chair proved to have some substance, since, in his opinion, the clients' representative from the CBI, the Forum's deputy chairman, Ian Reeves, put him under considerable pressure to alter the Board's constitution to give the private sector clients an even greater say in its affairs. As it happened, even had he agreed to do so, it was unlikely that the clients would have been able to exercise such a position by providing private sector client representation at the appropriate level.

Under Tony Jackson's chairmanship, the Board's activities expanded to cover an ever-widening area of the industry's affairs. From its establishment in February 1995 the Board met 11 times annually, only excepting the holiday month of August. During its lifetime (1995–2000), the Board took some 525 different papers at over 50 formal meetings, in addition to holding national consultative conferences and annual general meetings.

Inevitably the Board was criticised for importing an element of bureaucracy into its affairs. This was partly because of the difficulty of reconciling often conflicting interests of the different sectors represented on the Board and partly, in Don Ward's view, because the Department of the Environment, through its influential construction director, Phillip Ward, ensured that a civil service culture permeated the Board's administration. Tony Jackson expressed the view that the Board was always under-resourced, and that had the industry financed it to the extent of an annual £1m. subvention, with a corresponding Department of the Environment grant of £½m. annually, to support particular projects, this would have allowed the Board complete independence of the various industry sectors, and permitted it to impose improvements on the industry if necessary without having to seek consensus. Tony Jackson also felt that the main contractor sector, first through the CIEC and subsequently through its successor body the Construction Confederation, never gave its full-hearted commitment to the Board, but went along somewhat reluctantly without providing the full involvement of its member firms.

From the point of view of the clients, Stuart Humby, first CCF chairman and subsequently advisor to both the Treasury and the NatWest Bank, felt that the industry experienced a lack of leadership at the top, which both allowed vested interests to strangle initiatives at birth and generate a fragmented approach, making the industry's directing machinery, as personified by the Board, overmuch driven by a desire to achieve consensus.

Tony Jackson was succeeded in 1997 as chairman by Chris Vickers, who introduced a complete review of the Board's activities. These are described in detail in Chapters 8 and 9.

7 The development of the client movement

The client as leader of the process; definition of the client position; the preparation of the statement 'Constructing Improvement – the clients' pact with the industry'; input to the CIB; statement of research facilities and collaboration with the Construction Research and Innovation Strategy Panel (CRISP); publication of whole life costing document; launch of 'Constructing Improvement' at the Royal Institution; relationships with other industry sectors. Effectiveness. First suggestions of the Clients' Charter. The influence and implications of Private Finance Initiative (PFI) projects and Treasury/OGC policy on procurement.

The Construction Clients' Forum (CCF) ran from the end of 1994, when it was established as a direct result of the recommendation in Sir Michael Latham's report, to early 1999, when it transformed itself into the Confederation of Construction Clients (CCC). It can justifiably claim to be the first attempt comprehensively to bring together both public and private sector clients, to provide a single focus whereby clients could collectively influence the policy of the construction industry in the UK by reflecting client needs and implementing the client-led leadership of the industry which Latham had so strongly advocated.

In its early days, the Forum preserved the idea of a loosely organised grouping of clients, avoiding bureaucracy and formal structures. It concerned itself with articulating high-level professional issues, emerging as a result of exchange of information and experience between client organisations large enough, and with sufficient commercial influence, to force the supply side to take notice. This initial determination to minimise administrative formality dictated also a resolution to limit strictly the resources to be devoted to the establishment and running of the organisation. The clients involved in the initial establishment of the Forum were convinced of the need for the movement to be independent, and of the corresponding requirement therefore that its financing should be self-generated from subscriptions. As noted above (Chapters 4 and 5), the administrative support to the early Forum was provided by the part-time allocation of officers from the BPF and CIPS, both organisations which had been involved from the start in the Latham review, with the BPF providing office space and services at rates relatively low by the commercial levels of the time. Initial subscription rates were levied at £5k p.a. for the majority of the client organisations participating, which produced an operating budget of some £85k for the first year.

By 1996 the Forum had managed to enthuse an imposing representation from the client sectors which the supply side of the industry sought to woo, recognising as they did that their long-term commercial interests were likely to be best served by collaboration with them. The private sector was represented in the Forum by the BPF itself, the CBI, BAA, National Power, Northumbrian Water, and British Telecom. Universities and higher education were represented by Bristol University, on behalf of what was then the Committee of Vice Chancellors and Principals (CVCP), later to become Universities UK. Local authorities included Hertfordshire County Council on behalf of the Local Government Association, and for central Government the Central Unit on Procurement of the Treasury (later to become the Office of Government Commerce), the Highways Agency, MOD, NHS Estates and the Department of the Environment. Social housing was represented by the National Housing Federation.

Although this grouping looked impressive, as was noted in Chapter 4 there were serious gaps in its coverage of the various client sectors, sufficient to call into question the Forum's claim to be the comprehensive single-client voice that the industry sought. Several major clients in the private sector, notably those with predominantly civil engineering based programmes such as the Highways Agency, BAA, and Railtrack, and also Sainsbury's and Marks and Spencer among major retailers, while expressing support for the idea of a single client voice, made their contribution to the exchange of information through the informal grouping of private sector clients and suppliers in the Construction Round Table (CRT), and did not formally commit themselves to paying membership of the Forum.

The important offshore oil and gas construction clients, whose interests had earlier been defined, even before Sir Michael Latham undertook his review, by the CRINE initiative (Cost Reduction in the New Era), offered collaboration through the Capital Projects Clients' Group (CPCG), but this grouping was so loosely organised that it proved unable, or unwilling to commit its individual member firms to subscribe to the Forum.

Nonetheless, considering that no comprehensive client representative organisation existed before 1994, the CCF can point to considerable achievement in the four years of its life. The CCF, in collaboration with the industry:

- provided the client input to the production of the Construction Industry Board guidance documents and codes of practice, and represented the collective client voice in the industry's consultative machinery;
- produced elementary guidance to small and occasional clients on the construction process (*Thinking of Building*);
- embarked on preparation of guidance to clients on why whole life costing/performance should be the basis for construction procurement;
- established client panels which developed client-orientated strategies for research and innovation in the industry, and for contracts and contract management.

In all its activities the CCF consistently promoted the concept of, and training in, good clientship, defined as:

- clients raising their aspirations, skills and competence as buyers of construction;
- clients meeting their contractual obligation in a positive, fair and timely manner;
- clients incentivising and rewarding the supply chain for its contribution to delivering added value to clients;
- clients holding the supply chain accountable if and when it fails to deliver.

The approach by the Forum, incomplete though it may have been, convinced the supply side, represented in the Construction Industry Board, of the reality of the clients' commitment to achieve change and better performance in the industry by leading it to embrace good practice, and of the clients' resolution to improve their own performance. Hence, to overcome some suspicion, voiced from time to time in the Board and elsewhere in the construction press that the Forum appeared to be long on vague expressions of desire for improvement, but short on actual evidence of willingness itself to change its practices, the Forum embarked on a unique and important initiative to formulate a pact between clients and the industry. This was aimed at working collaboratively with all parts of the UK construction industry, within the Construction Industry Board and in other areas, to achieve value for money through best practice in construction.

Under the leadership of the Bursar of Bristol University, David Adamson, a member of the CCF Executive Committee and by then deputy chairman of the Forum, a working party began the preparation of a document setting out what the clients wanted from the industry, how they envisaged this might be achieved collaboratively, and committing themselves to identifying and implementing best practice in construction procurement. The working party consisted of:

David Adamson	Committee of Vice Chancellors and Principals; Chairman of the Working Group
Roger Aldridge	Marks and Spencer plc
Charles Botsford	Procurement Practice and Development, HM Treasury
Clive Cain	MOD (Defence Estate Organisation)
Frank Griffiths	Chartered Institute of Purchasing and Supply
John Hesp	Local Government Association
Stuart Humby	Chartered Institute of Purchasing and Supply
John Kerman	Highways Agency
Amanda McIntyre	Confederation of British Industry
Roy Morledge	Construction Procurement Research Unit, Nottingham Trent University
Rob Pearce	Marks and Spencer plc
Ian Reeves	Confederation of British Industry
Graham Robinson	Centre for Strategic Studies in Construction, Reading University
Geoffrey Wort	John Laing plc
Tony Pollington	Executive Secretary, CCF

with assistance from:

Ken Treadaway, Construction Round Table; Kevin Owen, Department of Surveying, Nottingham Trent University; Don Ward, chief executive, Construction Industry Board; Mike Burt,

HM Treasury; Richard John and Hywel Davies, Construction Research and Innovation Strategy Panel; Jeff Channing and colleagues, Construction Industry Sponsorship Division, Department of the Environment, Transport and the Regions (DETR); Richard Kauntze, British Property Federation; Colin Price, Northumbrian Water Ltd; and Geoffrey Wright, Hammerson plc.

In their introduction to the Pact, which appeared under the title of 'Constructing Improvement – The Clients' Pact with the Industry', the Forum members involved in its preparation said:

> CCF clients are concerned that although at its best the British construction industry is world class, too often its clients are let down by cost and time overruns, poorly performing technical solutions and contractual disputes. Until now clients have not been well enough organised collectively to call for better service from the industry, but now, through the CCF, they are able to do so.
>
> Recognition of the clients' strong influence, resulting from their purchasing power and the increased focus in business generally on customer requirements, allows them to press for rapid change in the industry to achieve the world class solutions that they seek. The clients recognise that to achieve change they need to develop the concept of 'good client-ship', and have identified the main areas in which improvements are needed. They want progress to be made by working with the supply side to adopt best business, design and construction practice rather than by relying solely on legislation, which may take considerable time to implement. The clients therefore welcome the industry's commitment to a Construction Best Practice Programme and endorse adoption of the Codes of Practice produced through the machinery of the Construction Industry Board. However, the clients see it as essential, if they are to maximise realisation of this opportunity for change, for their efforts to be complemented by various actions on the supply side. The clients represented on the CCF are therefore proposing a Pact to work with the industry to implement a plan of action to achieve the mutual benefits and the improved productivity which is within their grasp.

Under the terms of the Pact, clients represented on the Forum committed themselves to:

- set clearly defined and quantified objectives for each project and realistic targets of achieving it;
- pool information about construction and benchmark performance across the industry;
- communicate decisions quickly through the project sponsor;
- promote relationships based on teamwork and trust, and work jointly with all our partners to reduce costs;
- where unanticipated savings in project costs result from innovative thinking, in appropriate circumstances share them with the relevant parties;
- share experiences and information with the industry so that everyone can learn to undertake improved construction;
- appraise whole-life costs, not just the 'bottom-line';
- use client influence to improve statutory regulation where it is burdensome;
- support training and the improvement of standards;
- educate their own decision-makers in good client-ship;
- not unfairly exploit their purchasing power but look to form lasting relationships with the supply side;
- apportion risk sensibly in project contracts;
- improve their own management techniques and become better informed about the construction process.

They undertook to work together with the supply side of the industry to achieve change in specific areas:

- presenting clients with objective and appropriate advice on the options and choices to meet their needs;
- introducing a 'right first time' culture with the projects finished on time and to budget;
- eliminating waste, streamline processes and work towards continuous improvement;
- working towards standardisation in components where this provides efficiency gains;
- using a properly trained and certificated workforce and keeping skills up to date;
- improving management of supply chains;
- keeping abreast of changing technology by innovation and investment in research and development.

The Pact set out specific undertakings to be given in relation to:

- preparation and design;
- contracts and contract management;
- the industry's supply chain;
- standards of management and skills;
- quality from a client's point of view;
- avoidance of defects;
- research and innovation.

As evidence of the importance member clients attached to this statement of intent they said:

> The message from the Construction Clients' Forum is clear. Clients represented on the CCF will seek to place their £40bn of business with companies that are seen to follow the approach described in this document, and will seek such commitment prior to tendering, commensurate with relevant national, European and international regulations.

David Adamson and Geoffrey Wort discussed the draft in detail with Sir John Egan and Simon Murray, who were involved in the preparation of what became the 'Egan Report'; when the latter was launched there was comment how most of its points were similar to those in 'the Pact' – the chairman of CCF suggested that this should have been acknowledged.

The Pact was publicly launched at a prestigious gathering of construction industry and Government representatives on 19 March 1999 at the Royal Institution in Albemarle Street, Piccadilly. It received immense acclaim, both in the industry and in the media. In accordance with a policy adopted by the Forum at its inception, whereby chairmanship of the Forum alternated between private and public client sector representatives, Terry Rochester CB, recently retired chief highways engineer of the Highways Agency, had assumed the chairmanship. When he attended a meeting of the Construction Industry Board on the afternoon of the Pact's launch, he was greeted by the Board's chairman, Tony Jackson, with the words: 'You did us proud today.' This wholehearted welcome by the industry of the clients' commitment in the Pact was reflected in a joint letter to *The Times* from the chairmen of the supply side umbrella bodies (see Figure 7.1).

As further evidence of how well it was to be received, two major figures in the industry, Sir Martin Laing, president of the Construction Confederation (which had by then superseded the former CIEC

Construction Industry *Board*
The Building Centre 26 Store Street London WC1E 7BT   Tel 0171 636 2256   Fax 0171 637 2258

BUILDING A BETTER BRITAIN BETTER
(A STEP FORWARD FOR THE UK CONSTRUCTION INDUSTRY)

Sir,

On March 19th, clients representing some 80% of the UK construction industry's total turnover of £58 billion published documents calling for a new pact with consultants, contractors, specialists and product suppliers to improve the effectiveness of the building process and to contribute thereby to a significant, measurable improvement in the United Kingdom's economic performance.

Through the Construction Clients Forum, purchasers of buildings and infrastructure, from both private and public sectors, are setting out a comprehensive series of initiatives aimed at establishing good 'client-ship' and seeking joint action with the supply side of the industry to build a partnership where both sides are committed to continuous improvement through change and better practice.

This initiative is a central and fundamental aspect of the Latham Review (1994) which led to the creation of the Construction Industry Board, whose main principles, plainly stated, encompass fully meeting the needs and expectations of clients with the latter, in their turn, acting positively to make this happen.

The move both complements and adds to the value of the Agenda for Change published by the Construction Round Table, a group of influential major clients which is also pressing for higher standards and better performance.

As joint members of the Construction Industry Board with the Construction Clients Forum and the Government we would like to welcome the initiative of our colleagues and promise our total support.

Yours faithfully,

A J Jackson CBE. Chairman, Construction Industry Board
Chris Vickers, Chairman, Construction Industry Council
Sir Martin Laing CBE. Chairman, Construction Industry Employers Council
Peter Shiells, Chairman, Constructors Liaison Group
Peter Johnson. Chairman, Alliance of Construction Product Suppliers

Figure 7.1 Joint letter to *The Times* from the chairmen of the supply side of umbrella bodies.

Construction Industry Employers' Council), and Sir Nigel Mobbs, chairman of Slough Estates, contributed their own vision of a future for the industry based on implementation of the Pact, in a Foreword (see Figure 7.2).

Publication and acceptance of the Clients' Pact with the industry may well represent the high point of achievement of, as well as client support for, the client movement, certainly in its personification as the

FOREWORD

BY SIR MARTIN LAING AND SIR NIGEL MOBBS

Constructing Improvement is a most welcome addition to several publications and initiatives which explain how the client can obtain substantially better value for money. In return, the construction industry can expect improved profitability. These apparently irreconcilable aspirations are achievable by the collaborative and even-handed approach that this document advocates, but in order to bring about the improvement everyone is looking for, all parties involved will need to adapt.

In many ways the industry has changed remarkably in a few years; but how far does this change have to extend to meet clients' needs? If we were to take a view untrammelled by existing structures and constraints what might an ideal construction industry be like in twenty years?

▶ There would be an overriding priority for the industry to delight the client and understand how to contribute effectively to meeting his business needs in an environmentally benign way. But this could not be done without the positive involvement and commitment of clients.

▶ A much smaller pool of constructors and professionals to cover the main market area would be organised in alliances which would concentrate on delivering and continuously improving client value. To achieve this there would be a high degree of appropriate standardisation and whole buildings and other works would be "branded" with associated performance assurances.

▶ Management methods would have developed sufficiently to allow a high degree of certainty in the prediction of whole life costs, project delivery times, quality and safety performance and environmental impact.

▶ Finally, information technology would be all-pervading and first experiences of the industry by clients would be from a walk through a selection of building options using virtual reality. Not only would this overcome the difficulty of interpreting drawings but it would lead to really interactive design!

THIS IS A VISION OF THE FUTURE, BUT NOT A VISION WHICH CONFLICTS WITH A GROWING CONSENSUS ON HOW JOINTLY CLIENTS AND THE INDUSTRY CAN PROSPER AND IN DOING THIS CONTRIBUTE TO THE ENHANCEMENT OF THE BUILT ENVIRONMENT.

Sir Martin Laing CBE
President
CONSTRUCTION CONFEDERATION

Sir Nigel Mobbs
Chairman
SLOUGH ESTATES PLC

Figure 7.2 Foreword provided by Sir Martin Laing and Sir Nigel Mobbs to the Clients' Pact.

CCF. It remains as valid a statement today of informed construction client aspirations as when it was formulated in 1998/1999, and is a unique summary of aims and objectives which is still consulted today.

The subsequent approach to the implementation of the various programmes of action called for in the Pact proved somewhat less successful than could have been hoped for in the light of the euphoria engendered by the Pact's launch. A proposal by the Forum's new chairman, Terry Rochester, that separate client/supply side working parties should process programmes of improvement in the seven specific areas of undertaking set out in the Pact failed to find favour with all members of the Forum. There were increasing tensions in the CCF associated with what appeared to be a more dirigist chairmanship, and attendance at meetings fell steadily from about 30 when 'the Pact' was being finalised to a handful a year later. Some of the few remaining key players changed jobs.

It was decided to concentrate on the preparation of guidance for clients on the use of whole-life costing as the basis for construction procurement decision-making, and, with the support of DETR (formerly DOE's) Partners in Technology programme, Building Research Establishment and the Construction Round Table, this was published in 2000. The Forum also prepared a comprehensive statement on the priorities that clients would like to see adopted in the national construction research and innovation programme as overseen by the industry's research and innovation advisory body CRISP (Construction Research and Innovation Strategy Panel). This was widely welcomed by the research community, and the clients responded well by encouraging the membership of David Adamson and his move to the CRISP executive. The Forum also collaborated with DETR, and the Construction Best Practice Programme (CBPP) in inaugurating an annual survey of client satisfaction with the industry's standard of service, the first of which produced the rather odd finding that, while a majority of clients were dissatisfied with the quality of service received, 80 per cent of them would use the same supplier again. This, to some extent, reflected a continuing lack of expertise on behalf of clients, both in relation to their definition of what they really wanted, and in recognising the limited ability of the industry to deliver.

The achievement of the Forum in preparing, and obtaining acceptance of, the Pact should not however be underestimated. The welcome given to it was undoubtedly instrumental in the invitation issued by the new Secretary of State for the Environment and Deputy Prime Minister, John Prescott, at the 2000 Annual

Consultative Conference of the Construction Industry Board, to the clients to prepare a clients' charter, designed to implement programmes, and measurement of success, of improvements in the various stages of construction procurement. This challenge, accepted by the Forum's chairman, Terry Rochester on behalf of the clients, led eventually to the design and adoption of the Charter, now well-established and in use in some client sectors (see Chapter 9).

However, from the heights of the Pact and its follow-up in late 1999 and 2000, the client movement failed to build substantially on what had been achieved. With the conclusion of the initial 'Latham-influenced' phase of its activities, it was not entirely clear how the Forum could add further value for clients in its established mode of operation. The Construction Round Table (CRT), still active as an informal, predominantly private sector discussion club, despite a suggestion from its chairman, Martin Reynolds, construction director of Railtrack, that it should merge its secretariat with that of the CCF, had published its own Agenda for Change, largely anticipating the approach adopted in the Pact. The Government, evidencing some dissatisfaction with the rate of progress of improvement in the industry (see Chapter 8) and with the industry's consultative machinery in the Board, had instituted a further review under Sir John Egan, which had reported in July 1998 to the Deputy Prime Minister. There was a lack of clarity among member organisations of the CCF, and among client organisations not currently in membership of the Forum, about what precisely they required of a national client movement. Changing emphasis in the development of economic and social policies, including the recognition of the importance of an industry based on sustainable resources, suggested that a client movement needed to raise its perception of what is meant by a concept of good 'clientship', if it was to achieve the fundamental changes called for in the industry.

The CCF membership was virtually static, constituted predominantly of representative organisations rather than, as originally envisaged, actual contracting clients. It was difficult to define what needed to be done to attract a larger and more comprehensive membership. Despite the current client representation, the Forum was fundamentally under-resourced for its aspirations, let alone for any expansion of its influence. Changes in the industry, generated by the Movement for Innovation, the development of a Government Clients' Panel, likely change in central/local government relationships and expansion of the PFI approach to capital investment, all adversely affected the attitude of the industry and Government to the Forum's activities; CCF as a 'Latham' organisation was largely sidelined in civil service and political thinking. Hence, despite considerable goodwill towards the client movement generally, and acceptance by the industry of the Forum's sincerity, the Forum's status and standing was reduced, in that it was difficult, for example, to involve top level private and public sector decision-makers in its then current situation.

Accordingly, the CCF executive committee decided to undertake a fundamental review of the need for, and objectives of, a comprehensive client movement, and asked its deputy chairman, Ian Reeves who represented the CBI, to direct it. A consultant, Charles Botsford, of Botsford Consultants Ltd was appointed to support the review, together with the Forum's secretary, Tony Pollington. The review was directed by a steering group composed of:

Ian Reeves	Steering Group Chairman, Deputy Chairman CCF, CBI
Terry Rochester	Chairman CCF
David Adamson	Deputy Chairman CCF, CVCP
Martin Reynolds	Chairman CRT, Railtrack
(Ken Treadaway	CRT, alternate)
Douglas Weston	Millennium Commission

Ian Reeves adopted the following Terms of Reference:

> To advise members of the CCF, on what is required of a co-ordinated and/or integrated construction client movement, how the Forum can provide a mechanism for achieving this, and how the Forum's role and objectives might be developed to do this effectively.

> - To review what has been achieved to date and its effectiveness.
> - To consider what is required of a co-ordinated and/or integrated client movement in promoting the concept of good 'clientship' as the basis for achieving change in the industry, and how the Forum can provide the necessary leadership to achieve this in the context of present and forecast industry activity undertaken by central and local Government, other bodies such as GCCP, CRT, AGILE, CIB, Reading Construction Forum, CBPP, trade associations, professional institutes and others.

- To define members' expectations of a body which seeks to represent the totality of the construction client movement.
- To identify the type or types of membership, desired status, structure and organisation, funding, and objectives of any future Forum organisation, for it to operate effectively.
- To present draft conclusions and recommendations to the CCF Executive Committee, and final conclusions and recommendations to the full Forum.

He consulted widely, both within the Forum and outside, including the supply side, Government, and voluntary organisations concerned with the industry. He concluded that a radical approach to the future development of the client movement was necessary, both in relation to what it should do, and in relation to how this might be achieved through an appropriately structured client organisation.

His review raised a series of questions with clients and suppliers about the future position of the client relating to:

- clients' claim to leadership of industry;
- benefits to clients from client movement;
- benefits to clients' business from better client skills;
- collaboration with supply side;
- positioning of client in the industry;
- possible national structure for client organisation.

Consultation produced the following statement of objectives on behalf of clients:

- client aspirations demand strong client movement for success;
- supply side want empowerment from clients;
- clients must, in ultimate, be prepared to stand alone;
- actual contracting clients must be main drivers;
- objective of client movement must be added value to contracting clients;
- client movement must demonstrate progress by measurement;
- client must demand, and empower, innovative solutions;
- basis for buying construction must be whole-life value;
- clients wish to retain ability to network and share experiences and expertise;
- ultimate prize justifies combined investment.

The supply side emphasised that:

- suppliers want client leadership;
- they want assurance that clients favour quality rather than risk avoidance;
- they are dubious about the effectiveness of present industry structure;
- there was considerable goodwill towards clients' aspirations.

Ian Reeves' team recommended that the client movement should now concentrate on implementing the improvements identified as necessary for both clients and suppliers, and on ensuring that clients get best value from the various industry initiatives. They also emphasised that as presently constituted, the CCF had neither adequate resources, nor the comprehensive client coverage, to ensure that this could be achieved effectively and speedily. They drew attention to the fact that, while the total budget for 1999 for the CCF was £133k, of which £22k was required for the clients' contribution to the running of the Construction Industry Board, the comparable budgets for the CIB, the Construction Confederation, the Construction Industry Council, and the Association of Construction Product Suppliers (ACPS) were £250k, £8.5m, £1.5m and £700k respectively. The budgets of the Movement for Innovation and parallel organisations were vast.

Reeves' steering group recommended a client movement based on categories of membership, which emphasised that governance of the client organisation should be predominantly in the hands of large-scale client organisations, which actually undertook business with the supply side (see Figure 7.3):

Category 1 members: any client organisation able to demonstrate that it is an actual contracting organisation, with provision for specific groupings of:

- major clients (with a major rolling capital investment programme exceeding £100 million per annum averaged over the previous three years, but this figure can be reviewed);
- medium sized clients (with proportionate activity);
- small and occasional clients (ditto).

Category 2 members: sector representative bodies, including trade associations, able to speak on behalf of their client group.

Category 3 members: other bodies, including advisory groups and service organisations to clients who wish to be associated with the client movement.

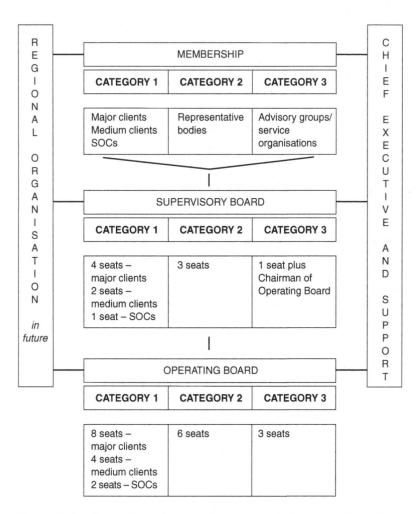

Figure 7.3 Ian Reeves' steering group's recommended structure for a client movement.

They calculated that their proposed structure would ensure annual core funding for the new organisation of approximately £750k.

The report and recommendations were presented to a major meeting of the Forum held at the CBI headquarters on 10 June 1999.

The Construction Round Table, which had supported the idea of a new style client organisation, made quite clear that, while it would work closely with any new organisation, it wished to retain its facility for managing confidential and informal discussions and exchanges of information among the organisations making up its original member- ship. Once this had been confirmed, the Forum gave general support to the Reeves recommendations, and action was put in hand for the launch of the new organisation, the Confederation of Construction Clients (CCC), to be chaired by a member of CRT, with an approach to procurement based on client to main contractor only ('one voice to one voice'), despite the strong contrary recommendations of Latham for a team approach.

The work involved in processing the new organisation, relating it to ongoing initiatives in other sectors of the industry and within the Construction Industry Board (CIB) are discussed in Chapters 8 and 9.

8 Change of government/ change of direction

The 1997 general election: pre-election preparation. Transition from DOE to DETR; general Government dissatisfaction with progress; decision to appoint Alan Crane; establishment of M4I; objectives, machinery for delivery; parallel existence/duplication between M4I and CIB; background to approaches to Sir John Egan; control of process by political advisors. Start of proliferation of Government-backed bodies, e.g. CRISP, M4I, Rethinking Construction, Housing Forum; discussion at CIB board meetings, particularly in July 1999; Labour Minister Nick Raynsford as president of CIB; CIB Millennium Conference 2000; adoption of targets; growing Government preoccupation with delivery of infrastructure improvements; Secretary of State John Prescott's challenge to the client body to produce a 'clients' charter'; development of the clients' charter.

Although there was evidence in the years 1995 and 1996 of a great deal of work being undertaken within the Construction Industry Board, and within its five component bodies (if the amount of papers, presentations, meetings and initiatives can rightly be construed as a measure of activity), there remained some intrinsic weaknesses in the strategic direction of the industry. The Board, although immensely active, was not fully viewed by the various sectors of the supply side as being really capable of eliminating the underlying suspicion and antagonism between main contractors, sub-contractors and specialist suppliers. The main contractors, by now consolidated into the Construction Confederation (CC), seemed luke-warm in their commitment to the future of the Board. Indeed, the Construction Confederation itself appeared divided on some issues. The different approach to construction business adopted by the large national and international construction companies, and the vast majority, some 80 per cent of construction firms in the UK, made up of small and medium sized firms led eventually to the major firms forming their own Major Contractors' Group within the Construction Confederation.

The inability of the clients' organisation to bring enough individual large private-sector clients within its membership left it vulnerable to criticism that it failed truly to represent the real influence of the value of private sector demand, as already noted. This was particularly true of important retailers, and also of the property developers, many of whose commercial schemes, particularly in the City of London, could be regarded as examples of client practice as then applied. Richard Saxon, then chairman, now a director of Building Design Partnership, who took a major part on behalf of the design professions in processing the recommendations of Constructing the Team, has indicated (letter 5 January 2004) that the client 'should be circumspect about developer involvement in reform. Only Sir Nigel Mobbs at Slough Estates seemed inspired by it'.

This apparent lack of conviction about the real prospects of fundamental and lasting reform resulting from the activities of the CIB and its component parts was reflected in the multifarious parallel activities, consultations and initiatives taking place contemporaneously with the Board's ongoing programme. This partly acknowledged that centralisation was an unpopular concept at the time, both in relation to the political climate and to the prevailing management culture. Any overt attempt by the Board to take an executive, or directing, approach to the way the industry undertook its business affairs would undoubtedly have proved counter-productive and unacceptable to the Board's members. Acceptance of the idea that the Board should take a co-ordinating position in an attempt to prevent duplication and wasted effort resulted only in the Board being consulted on various initiatives originating outside its own programme, with no guarantee that the Board could, or should, influence these.

Recognition of the risk of the Board proving only partially capable of achieving the dynamic pan-industry collaboration that was sought led to the emergence of a number of informal and formal groupings in the hope that these, released from the stultifying control of a formally structured, consensus-based body, could take quicker

and more decisive action. The Board's constituent bodies themselves
(the CC, CLG, CIC, ACPS, and CCF) established a group of chairmen and chief executives of these 'umbrella' bodies, ostensibly to
meet regularly to brief each other on the Board's agenda, but which
from time to time acted independently of the Board. The CLG saw
this as likely to weaken the CIB, which it consistently supported.
Because the Board was seen as 'apolitical', it was the Construction
Umbrella Bodies (CUB as it was known) that organised the industry's lobbying activities with central Government at the time of party
conferences. It was CUB also that organised a series of dinners in
Pall Mall clubs, at which ministers and senior civil servants met representatives of the industry, not all of whom were themselves
involved in the CIB's programmes.

Although the DOE's senior civil servant heading its Construction
Directorate, Phillip Ward over the Latham period, remained wholly
committed to the programmes established by the Board, and built on
the high level of respect and trust which he had established with all
parts of the industry, he found it necessary to be closely involved
with the Government's parallel and separate actions influencing the
industry. Some of these resulted from the success of the Treasury's
Central Unit on Procurement (later the Office of Government Commerce) in making Government aware of the importance of construction contracting in meeting Government's social and economic
objectives; some resulted from Government clients such as MOD
and the Highways Agency themselves reviewing their methods of
doing business in the light of the changing approach (for which the
reform movement can claim a modicum of credit), and some from
the efficiency scrutiny undertaken by Lord Levene's public sector
Efficiency Unit.

One of the major initiatives in the years after the Latham report
was the decision of the MOD, under the direction of the chief architect of its Defence Estates Organisation, Clive Cain, to seek to
remove the constraints on the encouragement of genuine
supplier/client partnerships in public sector work. This radical exper-
iment, eventually established in the public domain under the title of
'Building Down Barriers', was influential in the emergence of prime
contracting and framework agreements as a way of forming long-
term commercial relationships. The concept of 'target costing' as an
alternative to tendering in arriving at assessments of value and price
originated with 'Building Down Barriers' and was subsequently
adopted by another grouping of suppliers and clients, the Design
Build Foundation (DBF). At the same time as keeping in touch with

these, and other developments, including the Treasury's Efficiency Scrutiny and reviews of the CIB's activities undertaken by the National Audit Office and the Controller and Auditor-General, Phillip Ward had also to maintain contact with the Construction Round Table which continued to provide an alternative forum for clients and suppliers to get together.

Overshadowing all these activities at this time was the realisation that in the general election due in early 1997, on the basis of every opinion poll, a change of Government was likely. Given the commitment that had been made, and the input both in terms of intellectual effort and in the promotion of the new ideas consequent on acceptance of the Latham review, all sectors involved in the affairs of the Board were anxious to ensure that, whatever administration took over, the acceptance of the need for reform was maintained.

The Board therefore arranged for the Minister, Robert Jones, to attend their meeting in January 1997, at which the Conservative Government's commitment to achieving the improvements identified during the follow-up to the Latham report was confirmed. However, the original remit for the CIB was that it would continue in existence for a period of three years (1995–1998), and the Minister declined to commit any future administration to supporting the existing structure beyond that period. As was considered prudent at the time, given the indications of the public opinion polls, the Board then invited the opposition spokesman on construction, Nick Raynsford MP, to a subsequent meeting. Nick Raynsford had a long history of involvement in the affairs of the industry, primarily through his activities in London local government, and was both well known to, and respected by, senior figures in the industry. Although he was unable at that meeting to commit his party to a detailed line of policy in the event of them forming the next administration, he made clear that the reform of the industry was not a political matter, and that it was regarded as an important element in preparing the British economy for the twenty-first century.

The sense of impending change permeated both the affairs of the Board, and the perception of its constituent sectors of what might be expected for the future. Partly as a result of coincidence, and partly as a result of internal policy consideration, changes in crucial personnel took place immediately before, and immediately after, the general election. Six weeks before the general election the DOE announced that the head of its Construction Directorate, Phillip Ward, who had masterminded, and to a large extent directed, the follow-up to the Latham review would move within the Department to another directing post, outside the area of construction. This was a matter of widespread regret and press speculations in the industry, which had recognised Phillip's commitment to their efforts to improve, and had confidence in his willingness and ability to represent their interests within the Government machine. It was said that a move within the civil service was vital for him from a career development point of view, but several figures in the industry interpreted the Department's action as evidence of a civil service policy aimed at preventing any officer, however able, being perceived as being 'too close' to the interests of a particular industry, and therefore compromised in the advice they might give to ministers in different administrations. In an interview (9 July 2003), Phillip was at pains to deny that his move was anything other than part of a well-recognised civil service career path, but he accepted that the industry might have seen his move as reflecting a reduced commitment by the Department to the importance it gave to the industry in relation to national policy making.

Phillip was succeeded by John Hobson (a maths graduate from Sir Michael Latham's former college), who had considerable previous experience in Departmental/industry sponsorship roles, in water, waste management and particularly energy, where he had directed the Government's energy conservation programme.

In their contributions, both oral and written, to the compilation of this account, some of those who were involved at the time have voiced their strong conviction that John Hobson's brief was to sideline the programme of the Latham-generated Construction Industry Board, in view of the near certainty that a Labour Government would take over come the election, as indeed happened. John (interview 12 June 2003) has said firmly that he received no political steer, either before or after the election, which could be interpreted as seeking to belittle or diminish the Board or its constituent parts. He made clear, however, that the deputy secretary in the DOE with overall responsibility for overseeing this area of the Department's activities, Mavis MacDonald, pointedly asked what was the relevance of all the detailed work that had gone on, with major financial support from DOE, since the Board's establishment in 1995, in relation to the main conclusions and recommendations in the Latham review. John drew attention to the fact that there had been, from the start, considerable cross-party support for much of the programme, instancing particularly the all-party backing during the preparation and drafting of the Housing, Construction and Regeneration Bill. He maintains that, as far as DOE was concerned, the industry's future efficiency and effectiveness remained an important aspect of Departmental policy, and that they expected the in-coming minister to promote the need for urgent delivery of the targeted improvements, which the reform movement had influenced the industry to accept. This seems to be borne out by the opinion of Phillip Ward that in the handover period before John Hobson formally took over the Department's Construction Directorate, both he, and the Conservative Minister Robert Jones, made every effort to ensure that, subject to the agreement of the incoming Government, the movement towards change would be sustained. Robert Jones, as noted previously, had been active during his period as the minister in keeping his opposite number, Nick Raynsford, the Labour spokesman on construction, in touch with developments, and this reflects the general view that construction industry improvement was not seen as a controversial party-political issue.

As expected, the election returned a Labour Government with a very large majority, and within a week it was announced that John Prescott would become Deputy Prime Minister, with responsibility for a new, more powerful department, amalgamating the responsibilities for policies relating to the environment (including construction), transport and regional affairs (the Department for the Environment, Transport and the Regions (DETR)). As was widely expected, Nick Raynsford was appointed Minister with responsibility for construction and London and, of importance to the construction industry since it emphasised the importance the in-coming Government was to give to the Private Finance Initiative (PFI) and subsequently to the concept of Public Private Partnerships (PPP), Geoffrey Robinson became Paymaster-General.

While this was welcomed by the industry, the suspicion at the time was that the new Government might not give the industry the same level of consideration in relation to national policy as its predecessor. This was despite general recognition on the part of the industry that, if appointed, Nick Raynsford could prove an effective industry champion. An insider who was involved in the preparation of the general briefing for the incoming administration has subsequently indicated that this general brief extraordinarily made scant mention of the construction industry, and none at all of the CIB and its work, mentioning merely that the UK construction industry had been extensively reviewed (Latham), and that measures designed for its improvement were being developed. A view current among some of those involved at the time was that the incoming Government wished to claim credit for all the reform movement.

Recognising the importance of getting relationships with the new Government off on the right foot, the five umbrella bodies which made up the Construction Industry Board (but writing in their capacity as 'CUB' since it was not seen as appropriate for the Board to lobby in its own interests) wrote to the Deputy Prime Minister to press for continuing Government support for the Board's efforts (see Figure 8.1).

Britain needs building

26 Store Street, London WC1E 7BT.
Tel: 0171 323 3770. Fax: 0171 323 0307

13 May 1997

Rt. Hon. John Prescott MP
Deputy Prime Minister and
Secretary of State for the Environment, Transport & the Regions
Eland House
Bressenden Place
London
SW1E 5DU

Dear Deputy Prime Minister,

We, the chairmen of the five "umbrella" organisations which represent the whole of the Construction Industry, congratulate you on your new appointment. We all look forward to working closely with you.

There is much to be done to improve the built environment, especially the stock of buildings, houses - schools and hospitals - and we also firmly believe that the country needs a proper detailed transport strategy if we are to be competitive in Europe in particular. As we are sure you are aware, Sir Michael Latham's review, *Constructing The Team*, held out the promise of a 30% reduction in real construction costs; such a reduction represents many hospitals, schools and houses. A start has been made, led by the Construction Industry Board, and we very much hope that the Government will support this body and continue to play an active role in promoting best practice.

Construction is an important part of the economy with a total output five times that of Agriculture and greater than any single manufacturing industry. We are vitally interested in training and in the environment in general.

We would like to explore with you the ways in which we can help to meet Government objectives and to satisfy the needs of the nation. We would welcome an early meeting with you.

Yours sincerely,

Robert Napier	Geoffrey Wright	Christopher Vickers	Sir Martin Laing CBE DL	Allan McDougall
Chairman	Chairman	Chairman	Chairman	Chairman
Alliance of	Construction	Construction	Construction Industry	Constructors
Construction	Clients Forum	Industry Council	Employers Council	Liaison Group
Product				
Suppliers				

Figure 8.1 Letter to John Prescott, DPM and Secretary of State for the Environment, Transport and the Regions, from the Construction Industry Board, pressing for continued government support for the CIB.

As noted in Chapter 7, very soon after the election illness forced the resignation of Sir Ian Dixon as CIB chairman. After considerable difficulty in identifying a successor, Tony Jackson became chairman of CIB, with the then chairman of the clients' organisation, the CCF, Geoff Wright, moving to become deputy chairman of CIB. This in turn required the CCF to appoint a new chairman, a procedure which did not prove easy. The convention, by then established within the CCF, called for the chairmanship to alternate between representatives of public and private sector clients. Chairmen up to this time (Stuart Humby from NatWest and Geoff Wright from Hammerson's) had come from the private sector, and candidates from the public sector were identified as David Adamson (University of Bristol representing the university/higher education sector) and Terry Rochester (retiring from the Highways Agency, representing the central Government client sector). However, as is often the case in quasi-voluntary bodies made up of potentially disparate interests, there were internal pressures within the client movement to bypass the convention and retain a private sector chairman. The candidate was Ian Reeves from the CBI, but his position was far from being universally accepted by both clients and the supply side, since he was seen in some quarters not as a genuine contract-signing client, but as a consultant client adviser. At the same time, the main contractors on the supply side indicated that they would be happier with a central Government client sector chairman, if it was inevitable that a public sector candidate would be chosen. The choice fell on Terry Rochester, who had sufficient time for the appointment; the lack of unanimity, however, in the choice of candidate undoubtedly left a residue of mistrust in the movement.

Hence there was a situation whereby not only was there a change in the political area with a new Government and a new minister, but there were also new chairmen of the CIB and of the client organisation. Whether as a consequence, or whether the new Government had previously decided that progress in the Board's programme was too slow, the perception emerged that the existing strategic development of the industry, and the organisational structure to implement it, did not enjoy the full support of ministers and the Department. This was compounded when, at the first annual general meeting of the Construction Industry Board, which took place within a week of Tony Jackson assuming the Board chairmanship, the new Minister Nick Raynsford notably omitted any mention of the Board, its programme, or its constituents in his keynote address. Instead he concentrated on the new Government's determination to push forward major change in various sectors of the national economy through the establishment of task forces, the number of which eventually exceeded 100. He announced that the Deputy Prime Minister and Secretary of State for the Environment, Transport and the Regions had asked Sir John Egan, then chairman of BAA to chair a Construction Task Force to 'improve the quality and efficiency of UK construction'.

Those currently involved in the consultative machinery of the industry, and those who were in leading positions in the Board and who have contributed their personal opinions to this account of what went on at the time, have maintained that this establishment of a task force to effect improvement in the industry should not be regarded as evidence of Government dissatisfaction with the activities of the Board and the industry sectors, including the clients, involved in it. On the other hand, others from outside the immediate circles of the Board and the Department, remain convinced that this was the case.

Sir John Egan's terms of reference were:

> To advise the Deputy Prime Minister from the clients' perspective on the opportunities to improve the efficiency and quality of delivery of UK construction, to reinforce the impetus for change and to make the industry more responsive to customer needs.

The Task Force will:

- quantify the scope for improving construction efficiency and derive relevant quality and efficiency targets and performance measures which might be adopted by UK construction;
- examine current practice and the scope for improving it by innovation in products and processes;
- identify specific actions and good practice which would help achieve more efficient construction in terms of quality and customer satisfaction, timeliness in delivery and value for money;
- identify projects to help demonstrate the improvements that can be achieved through the application of best practice.

> The Deputy Prime Minister wishes especially to be advised on improving the quality and efficiency of housebuilding.

These terms of reference (excluding the reference to house building) closely resembled those given to Sir Michael Latham some four or five years previously. The members of the Construction Task Force were:

Chairman: Sir John Egan (Chief Executive, BAA plc)
Mike Raycraft (Property Services Director, Tesco Stores Ltd)
Ian Gibson (Managing Director, Nissan UK Ltd)
Sir Brian Moffatt (Chief Executive, British Steel plc)
Alan Parker (Managing Director, Whitbread Hotels)
Anthony Mayer (Chief Executive, Housing Corporation)
Sir Nigel Mobbs (Chairman, Slough Estates and Chief Executive, Bovis Homes)
Professor Daniel Jones (Director of the Lean Enterprise Centre, Cardiff Business School)
David Gye (Director, Morgan Stanley & Co Ltd)
David Warburton (GMB Union)

None of these were members of the CIB, or had had any input into its affairs. Sir John's appointment seems to have followed an approach by the Deputy Prime Minister (DPM) to the Chancellor for an allocation of some £5bn to 'pump-prime' the programme of infrastructure investment to achieve the public service improvements promised in the Labour manifesto. Chapter 10 notes a pre-election dinner party between John Prescott, Gordon Brown, Peter Mandelson and John Egan where the 'pump-priming' was agreed. At a subsequent meeting between the DPM, the Paymaster-General and Sir John Egan (who, as chairman of BAA and an experienced and articulate client of the industry albeit an extremely critical one, had been invited by the DPM to 'do something' about construction) it was, however, made conditional on evidence of the industry's commitment to change. In the light of Sir John's indication of his willingness to take the lead, the message was sent from the DPM's office to the Department to give the fullest support, including financial, to the establishment and speedy conclusion of the Task Force's programme of work.

The Task Force's work was largely undertaken by Simon Murray, on the staff at the time of BAA and a close associate of Sir John Egan, and by other staff of private sector firms represented in the Construction Round Table. Consultation with the existing representative bodies in the industry was purposely kept at a minimum, although following strong representation by Tony Jackson, chairman of the Construction Industry Board, a CIB 'shadow Egan group' was established to ensure that the new investigation was at least aware of what had been, and was being, done already. However positively this may be represented, it is hard to conclude other than that the Construction Task Force was unimpressed with the achievements so far and felt that a more dynamic approach was necessary.

The Task Force's conclusions emerged in July 1998 as Rethinking Construction. Its recommendations and proposals for action are discussed in Chapter 10. There were recommendations to expand the Construction Best Practice Programme into an industry 'knowledge centre', and for the establishment of a comprehensive and sophisticated programme of demonstration projects. Other recommendations build upon those of the earlier Latham review and the clients' 'pact with the industry'. The considerable body of work, events and achievements of the Egan elements of the decade of reform are set out at greater length in Chapter 10.

Interpretation of the reasons for, and results of, the establishment of the Egan Task Force are varied. Questioned about his views as to whether or not the incoming Labour Government in 1997 had decided to follow the industry-wide collaborative effort in implementing his recommendations in *Constructing the Team*, evidenced through the activities first of the Review Implementation Forum and subsequently through the more structured approach of the Construction Industry Board, Sir Michael Latham did not feel that this caused a problem. He emphasised that he had always enjoyed good personal and professional relationships with Nick Raynsford, the in-coming construction minister, who continued to consult him on professional matters connected with the industry after taking up office. Michael expresses considerable personal satisfaction at the continuity that this open-handed approach by Nick Raynsford established. On the other hand, he expressed strong disappointment that the new Secretary of State, John Prescott, evinced little interest in the industry's efforts to reform itself, and resolutely avoided meeting him or indeed acknowledging the previous work that he had done on the review. Michael Latham put this down to the Secretary of State's refusal to see the industry as important in relation to the national economy, or to the implementation of Labour's election promises on the public services, and reflecting ideological suspicion of his (Michael Latham's) previous Conservative affiliations. Michael Latham stressed that the decision to ask Sir John Egan to take forward the reform movement with added momentum stemmed not so much from Raynsford's and Mavis MacDonald's fear that the movement had run into the sand, as from John Prescott's personal contact with Sir John Egan. John

Prescott regarded him as a dynamic businessman with strong Labour Party sympathies, whose record as a construction industry client during his chairmanship of BAA convinced him that he could bring together a strong group of individual private sector clients to drive the suppliers to early adoption of a more modern and effective way of doing business. Sir Michael pointed out that he personally had no difficulty with the approach adopted by Sir John Egan, whose recommendations in *Rethinking Construction* largely reflected his own in *Constructing the Team*. He emphasised that at no time had he ever criticised the Egan report, despite a quite widely held view across the industry that much of it reflected what had already been published in *Constructing the Team* and elsewhere. He noted that Egan was widely respected by the industry as a strong and involved client, but thought that his approach to the industry was itself somewhat antagonistic; and that this inhibited the industry at large from promoting the collaborative environment that seemed to be there for the taking after publication of *Constructing the Team*. He summed up the position post-Egan as being one where the industry respected John Egan but did not hold him in affection; this was, in Michael Latham's view, evidenced by the industry's reluctance to involve him in industry functions, whereas Michael Latham found it difficult to decide between the myriad invitations to address the industry representative bodies and other organisations connected with the industry. Sir Michael Latham did acknowledge that the reform movement experienced a downbeat period after the change of Government in 1997, but attributed this more to the effect of changes in personnel involved rather than to any pre-arranged change of policy.

In July 1998 Sir John Egan and Simon Murray came to part of a CIB meeting to announce publication of the Egan report. David Adamson recalls Sir John hammering most of the elements in the industry, starting with architects and quantity surveyors. He said he never remembered seeing so many senior people get so angry so quickly, but recalls reflecting that if this injection of energy, political support and resourcing could somehow work with and alongside implementation of the more cerebral Latham analysis, then the industry really would be on a roll. That this did not happen was, to him, a matter of lasting regret.

The CIB and the constituent bodies represented on it continued to work within the programme on completing the 12 guidance documents (see Chapter 5) initiated after the Latham review had been accepted in 1995. Tony Jackson, the new chairman of CIB, in particular retained a vision of CIB as the single strategic consultative body for the industry. Significant progress was made in CIB developing key performance indicators for the industry, a concept which had originated within the client body of the CCF, and which was adopted by the DETR as a policy priority. It led eventually to the design of a comprehensive 'spider's web' or 'radar' wallchart, developed by the Construction Directorate of DETR to a concept suggested by the CIB office, which oversaw its publication and distribution. Don Ward, the director of CIB, in retrospect sees this performance-assessment approach as one of the major and lasting achievements of the CIB, giving individual construction companies, professional advisors, specialist suppliers and materials and component manufacturers a useful tool for measuring their own performance over time, and relating it to the general objective of a 30 per cent productivity improvement over the five year period 1995–2000.

It is, however, clear that despite the incoming Government's declared reluctance to take the leading role in the industry's strategic development, preferring to see the industry itself provide the momentum for change, the influence of the Board and its 'umbrella' bodies began to decline after the election. The remit given to Sir John Egan to galvanise the industry to action, and the poor opinion expressed by John Hobson of the effectiveness of the industry's own representative bodies (including the Board) actually to achieve change led the relevant Government departments – DETR, OGC, and Treasury – increasingly to define policy objectives and design structures and machinery to achieve these. The Department began to draw increasingly on the input and expertise of figures in the industry unconnected with the earlier activities of the Board, but who had worked with and advised Labour politicians in central and local government. Besides Sir John Egan himself, Alan Crane, chairman of Christiani and Neilsen and active in the Labour group on Southwark Council, took over the chair of the steering group given the task, with Departmental financial backing, of finalising a set of key performance indicators for a series of demonstration projects, aiming to extend the original CIB project framework to whole-life performance of facilities in use. By the early part of 1998, three major developments outside the previous industry strategic machinery as personified by the Board had been established.

Under the chairmanship of the Minister, Nick Raynsford, with the Financial Secretary to the Treasury, Geoffrey Robinson, Sir John Egan and the CIB chairman, Tony Jackson, a Movement for Innovation (M4I) was inaugurated to work in five main areas of performance improvement; key performance indicators including demonstration projects, training/education/research chaired by Robin Nicholson from architects Cullinan and chair of the Construction Industry Council, culture change chaired by Stella Littlewood (ARUP), design and development of a knowledge centre (chaired by Ian McPherson from MACE Consultancy) and client/supply side relations (chaired by Shona Hay of AMEY).

A Construction Best Practice Programme (CBPP) was launched with major financial backing of some £2m p.a. from the Department, to provide information and help, structured around improvement themes or 'levers for change': lean construction; partnering and team choice and development; briefing the team; forms of contract and choice of procurement method; value management; sustainable construction; standardisation and prefabrication; benchmarking; integrating design and construction; supply chain management; risk management; culture/people issues; and information technology. Definitions, background information, summaries, business cases, and key references on all these levers were made through a website. Other services included an information line, IUKE company visits and workshops (delivered through an alliance with the CPN), and access to case studies. The programme was located in the Watford headquarters of the Building Research Establishment, and directed by Zara Lamont, on secondment to the Government from Carillion plc.

These major initiatives, originating from Sir John Egan's *Rethinking Construction* report and developed within a general programme of 'Accelerating Change', which he oversaw in collaboration with John Hobson in DETR's Construction Directorate, were designed to add momentum to the flagging programme of the CIB. They were promoted, however, very much in line with Sir John Egan's choice of energetic innovators in the industry, both suppliers and clients, known to him personally. With John Hobson, and Simon Murray from his own company, he drew particularly on active participants in the Construction Round Table, which, while being understandable given his commitment to quick action, detracted from the reputation of the CCF and their client representation.

The Board was anxious to collaborate with *Rethinking Construction* and with its spin-offs M4I and CBPP, but was inhibited by a notable lack of resources in comparison with those made available, mostly from public funds, by DETR to Egan. Tony Jackson, as chairman of CIB, obtained Departmental backing for the appointment of an experienced contract specialist, Tony Kemp, to provide a close relationship with the Egan group, but this had only limited advantage because of the latter's unavailability on a continuing basis.

In an attempt to retain a co-ordinating role for the Board an 'Egan group' was established within the CIB to provide a bridge from M4I to the wider industry as represented by the Board and to provide advice on implementation of the Egan agenda. At the same time, and due to the Government's acknowledgement that it was essential to bring the house-building industry into the general movement for improvement, a Housing Forum was established, with formal links to the Rethinking Construction organisation.

The result was the start of an increasingly complex consultative machinery, with considerable scope for the blurring of objectives, duplication of effort, obscure accountability, and suspicion of motives of different interests, as is illustrated in Figure 8.2. This structure was adopted and promulgated by the Board midway through 1998. The adverse and ultimately destructive effect of this on the CIB and its constituent parts is discussed in Chapter 9.

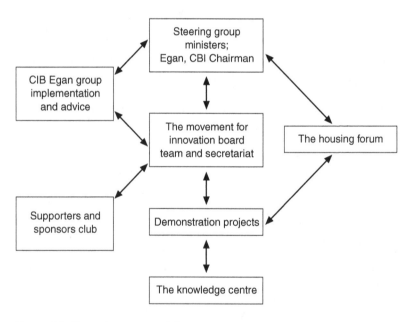

Figure 8.2 Central structure of the movement.

9 The beginnings of the dissolution of the Latham bodies

Report on CCF; proliferation of Government backed bodies – CRISP, M4I, etc.; discussions in CIB, particularly CIB board meeting July 1999; Raynsford as president of CIB; development of proposals for CCC; CIB 'Millennium Conference' 2000; growing Government preoccupation with delivery of infrastructure improvements; Secretary of State's invitation to client body to produce a client's charter; launch of CCC and clients' charter; NAO report on Egan, M4I etc.; new CIB Chairman Chris Vickers; first ideas about review of CIB; 'spoiling' tactics of CCC and disastrous long-term effects on CIB, collapse of CIB and CCC; departure of Egan.

Throughout 1998 emphasis shifted from the Construction Industry Board's key role, as originally perceived as providing a forum for liaison between the umbrella bodies of the supply and demand side of the industry and the Government, towards the action-based programme recommended by the Egan report Rethinking Construction. Tony Jackson, chairman of the Board, jointly with the Department's representative John Hobson chaired a major industry conference to establish formally the mechanisms for operating the activities called for in Rethinking Construction, particularly the Movement for Innovation with its 82 demonstration projects and various initiatives, the Industry Knowledge Centre, and the Construction Best Practice Programme, with its own demonstration projects.

In September 1998 the Board recommended, and DETR agreed to, a new strategic framework for its activities for a further two years and in November 1998 adopted an operational plan for the next year

(1999). This identified CIB's core responsibilities (in addition to its liaison role referred to above) as:

- good practice, particularly in procurement;
- productivity and cost improvement with an emphasis on enhancing value for clients at all levels in the supply chain;
- research, development, and innovation, particularly in support of improved procurement and processes;
- the public perception of construction;
- the use of statistics to provide common performance indicators to measure and inform pan-industry improvement.

As part of this overall operational plan, the CIB had set up a series of panels covering image, good practice, productivity and cost improvement, liaison with the Egan Task Force, performance indicators, and research and innovation. The CIB and its members contributed to the organisation and running of the Considerate Contractors Scheme (subsequently taken over by the contractors' umbrella body the Construction Confederation, the CIC and the CPA, acting together as CUB (Holdings) Ltd), National Construction Week, and Constructionline, a seven-year concessionary contract run by Capita on behalf of DETR to provide a suppliers' assessment system for use by public and private sector clients. In addition, the CIB marketed its practice guidelines on a commercial basis, through a limited company, CIB Ltd. It continued to sell copies of its codes of practice. These numerous and varied activities resulted in a complex organisational structure (see Figure 9.1).

At the same time, the client organisation, the CCF, was beginning its review of its role and organisation under the CBI's representative, and CCF deputy chairman, Ian Reeves, which would lead to recommendations in June of the following year (1999) for the launch of a new and, it was hoped, expanded client body, the Confederation of Construction Clients (see Chapter 1).

Despite the adoption of the 1999 Operational Plan, the relative success of the 1998 National Construction Week, the publication of the Board's guidance documents and continuing work on the definition and development of key performance indicators (KPIs), the feeling was growing that the Board, and its constituent parts were not achieving the rapid change in the industry that both Government and the clients of the industry sought. The Government was becoming increasingly concerned that the industry was still perceived by the general public as being inefficient, excessively costly, and produc-

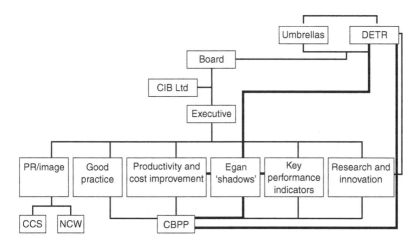

Figure 9.1 The complex structure of CUB (Holdings) Ltd.

ing poor quality projects. Horror stories in the media, many of them justified, about small and occasional clients, and to some extent large-scale clients as well, being ripped off by 'cowboy builders' (later referred to more officially at the behest of the supply side as 'rogue traders') were seen by the Government as potentially politically damaging. The Department, at the specific request of the Minister, established a Rogue Traders Working Party which, although chaired by Tony Merricks, at the time one of the deputy chairmen of the Board, conducted its work largely outside the Board's own programme. Nonetheless, the Board and its members contributed to the working party's report and considered offering to run an Accreditation and Consumer Guarantee scheme on a third-party contracted basis. Because of the relatively low level of resources available to the Board, this proposal was rapidly dropped.

The Department's Construction Director, John Hobson, continued to be in contact with the Board and its work, but it became increasingly obvious that he saw more immediate advantage to the Government and to the Department occurring through the expansion of M4I and Rethinking Construction. There was a widespread impression that he avoided publicly praising CIB or CCF until much later, and then briefly and under pressure. Research and development activities, for which the Department controlled financial allocations amounting to some £20m a year, had assumed considerably greater importance with the decision to privatise the Building

Research Establishment, and phase out, over a period of years, the number and value of Government-awarded research and development contracts. The Board made a contribution to the strategic definition of where research and development emphasis should be placed through the Construction Research and Innovation Strategy Panel (CRISP) which reported to the Board, but the real influence rested with the Department as paymaster. Don Ward, the Board's chief executive, has indicated that in his opinion the inability of the Board to control, or at least strongly influence, the placing of major research funding contributed to its weak position.

Throughout this period the calls for greater efficiency and professionalism among clients of the industry, articulated first in the Latham review, and given further backing in the commercial world by Sir John Egan's Rethinking Construction report, were becoming more widely accepted. This was particularly true in the public sector, where a report on *Construction Procurement in Government* by Sir Peter Levene's (now Lord Levene) Efficiency Unit in the Cabinet Office led to the production of a series of guidance notes relevant to works projects by the Central Unit on Procurement (CUP) in HM Treasury. CUP established a Government Construction Client Panel (GCCP) which issued 12 comprehensive guidance notes over a period of some three years, aimed at translating the recommendations made in the *Efficiency Unit Report*, and those made direct to CUP by consultants Bath University School of Management Agile Construction Initiative, into practical proposals for implementation by central Government departments. It was the hope, only partially realised, that this newfound emphasis on the importance of good procurement practice would in due course be adapted for use in other parts of the public sector, particularly in local government. These guidance notes were produced within a structured framework (see Figure 9.2) and were seen as part of a wider concentration within the Government service, promulgated by CUP, on improving public sector procurement generally.

The determination of the Treasury's Central Unit on Procurement (later to become a Government Office in its own right as the Office of Government Commerce) to promote the new concepts of construction procurement and contracting among central Government departments as clients cannot be over-estimated. Echoing the findings and recommendations of the Latham review, CUP and later the OGC embarked on and carried through a programme of modernising construction procurement practice in central Government which led ultimately to the adoption by some Government clients of

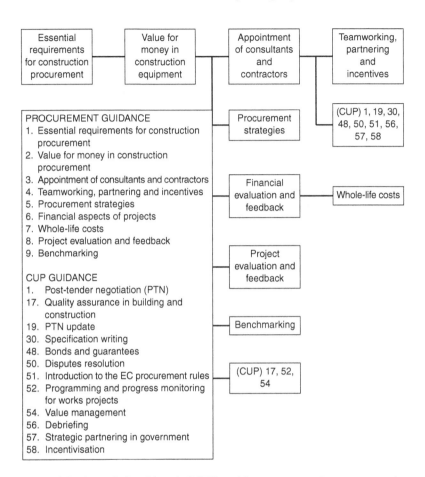

Figure 9.2 The relationship of GCCP guidance notes relevant to works projects.

high standard practice aimed at achieving excellence. Central Government did not shirk from subjecting itself to vigorous self-examination. Following the publication of the Latham review in 1994, Sir Peter (later Lord) Levene led the Efficiency Scrutiny of Government construction procurement in 1995. Following publication of Sir John Egan's report in 1998, Sir Peter Gershon, the head of the newly developed OGC, conducted his own review as part of his general examination of Government procurement, and the National Audit Office undertook an audit-based examination itself in 2001. As part of earlier reviews, the Treasury commissioned benchmarking studies through Bath University in 1998 and 1999. These showed that 73 per cent of Government projects exceeded the tendered price in 1998, and 70 per cent exceeded the time estimate at the award of contract. By 2005, these excesses had been reduced to 50 per cent and 66 per cent respectively, a significant improvement, albeit still a relatively unimpressive performance compared to the efficiency of other UK industries.

The various internal Government reviews referred to above showed the problem among Government clients to be the same as that facing clients generally:

- adversarial approach;
- lack of integration;
- poor collaboration in problem solving;
- short-term relationships;
- lack of supply chain management;
- concentration on lowest cost irrespective of quality;
- poor health and safety record;
- lack of learning culture;
- poor customer focus.

The OGC identified the need for better performance for construction to meet user requirements, lower whole-life and operational costs, greater cost and time predictability and the elimination of waste. In 1999 they launched a three-year action plan focusing on:

- management and culture change;
- measurement;
- standardisation;
- integration.

Emphasis was placed on the use of risk and value management, whole-life costing, post-project implementation, client performance, and measurement of success through performance indicators. A focus on integration concentrated on procurement strategies, particularly design and build, PFI, and prime contracting, with emphasis on teamworking and partnering. Government policy on sustainability led to the adoption in July 2003 of targets to be achieved by March 2005 in relation to:

- whole-life costing;
- designing out waste;

- alternatives to new build;
- energy efficiency;
- environmental impact.

OGC also set strategic targets to be achieved by March 2005 to reduce average procurement time for Government projects from start of procurement to award of contract by between 15 and 25 per cent, depending on the type of project, and for 70 per cent of construction projects reaching a stage enabling their prospective benefits to be evaluated in the period April 2003 to March 2005 to be delivered on time, within budget, with zero defects and meeting users' expectations.

The original GCCP Guidance Notes (see above) were consolidated into 11 procurement guides under the general title of 'Achieving Excellence in Construction', covering:

- initiatives into action;
- project organisation;
- project procurement life-cycle;
- risk and value management;
- integration of the project team;
- procurement and contract strategies;
- whole-life costing;
- improving performance;
- design quality;
- health and safety;
- sustainability.

These have been widely used throughout the public sector.

Meanwhile, work continued within the client body, the CCF, on their review of the future requirements for, and organisation of, a construction client movement. Particular emphasis was given in the review of CCF by its deputy chairman, Ian Reeves, to the importance of finding ways of involving significant private sector clients in what had by now come to be recognised as the construction industry reform movement, and his proposals (see Chapter 7) for an entirely new client body were eventually accepted by the CCF, and formally launched in July 1999 (see below).

However, the unease felt in the Construction Industry Board about the true level of support offered to it by the supply side, particularly the main contractors, the lukewarm endorsement given to the Board by DETR's construction director, John Hobson, and his concentration on the Egan task force and M4I developments, and a growing feeling within the group of private sector clients working within the Construction Round Table that the Board was proving ineffective in achieving a speedy move towards a customer-based industry, convinced the Board chairman, Tony Jackson, that a review of the CIB was urgently necessary.

As earlier described, the Board had already undertaken an investigation into its future beyond 1998 in August of that year, and this had led to the adoption of an operational plan for 1999. This further review was proposed to be more fundamental, and involved seeking the views of all constituent parts of the Board, about what should be covered in this comprehensive examination of the Board's activities. The decision to ask Chris Vickers, the deputy chairman of the CIB who had been designated to take over from Tony Jackson, whose two-year period of CIB chairmanship expired in July of that year (1999), was announced at much the same time as an announcement that the results of Ian Reeves' review of the client body, the CCF, and proposals to change it into a new Confederation of Construction Clients (CCC), would be launched in the following year. These frequent announcements of internal reviews of various constituent parts of the industry, and of its strategic body, the Board, gave the unfortunate impression of a group of partially representative industry organisations bogged down in considerations of structure and administration, rather than in activities aimed at speedily achieving the defined objectives of Latham and Egan. This, together with an apparent lack of Government support, was brought into stronger relief by the vigorous and well-resourced parallel programmes of other organisations without formal relationship with the Board. As an example, an Urban Task Force had been established under the renowned architect and urban planner Lord Rogers, a recognised Labour supporter, and this task force published detailed recommendations to Government in summer 1999 calling for an urban renaissance, and a national strategy of compact urban development to meet the twin necessities of solving the crisis in cities (serious unrest was evident in some decaying city centres at the time), and providing millions of new dwellings. The report highlighted the task force's concentration on the main instruments to achieve these objectives – land use planning, land assembly, urban design and finance, and targeting public and private investment. Many of the considerations involved were closely connected with what many saw as CIB's similar concerns, although it was not formally involved. While the Board could not claim that this work, nor that relating to the development of the house-building industry undertaken by the recently established Housing Forum, should come entirely within the Board's overall remit (and indeed the Government would not have accepted such a transfer of responsibility for public policy from the relevant Department, the DETR), taken with the powerful and widely publicised activities of Rethinking Construction and M4I, it gave the impression of increasing fragmentation, as well as sidelining the Board. 'The caravan had moved on.'

In June 1999, the Minister, Nick Raynsford, addressed the CIB's Annual Consultative Conference in his capacity as CIB president. In a wide-ranging speech he reviewed all that was going on, and emphasised that it was seen as a coherent programme of industry improvement to which the Government was fully committed.

He re-stated the Government objectives:

> First, to put the interests of individual clients and customers at the heart of the agenda. As with all other industries, responsiveness to the customer is crucial to success.
>
> Second, to encourage respect for, and, where necessary, protect, the wider community, the environment, future generations. And I include in that the promotion of good attractive, design: design that combines high aesthetic quality, fitness for purpose and sustainability.
>
> Third, to encourage respect for, and, where necessary, protect, those who work in the industry. No industry can afford to ignore the interests of those whose talents and skills design and deliver its products.
>
> Fourth, to support, and indeed to drive forward, investment in the modernisation of Britain by promoting a modern, efficient, productive, innovative, profitable construction industry that has the talent, the structure, the drive and the competitiveness to be a leader in an increasingly global market: a market in which truly multinational companies will play an increasingly important role.

Emphasis was given in his speech to Government-inspired initiatives such as Constructionline and the Quality Mark Scheme proposed by Tony Merrick's Rogue Traders (Cowboy Builders) Working Group, environment, training and design, and the appointment of Sir Martin Laing to head another high-level industry focus group to address the questions relating to sustainability. He concentrated heavily on the follow-up to Sir John Egan's Rethinking Construction, and the machinery established within M4I to achieve it:

- an M4I board, chaired by Alan Crane;
- a Housing Forum chaired by Sir Michael Pickard;
- a central Government task force chaired by Steve Robson, at the time a Second Permanent Secretary at the Treasury, bringing together chief executives of Governmental agencies and senior departmental officials together responsible for 70 per cent of central Government's construction procurement;
- a local government task force for construction to be established.

Reflecting a view increasingly widespread about 'initiative fatigue', the Minister recognised the risk of duplication of effort that might result from all these initiatives, and the difficulties inherent in them for the CIB. He said:

I recognise the concern that we are in danger of creating too many separate overlapping initiatives and organisations. To some degree a broad spread is needed so as to involve as wide a spectrum of the industry and clients as we can. But once we have the movement up and running, we will need to ensure effective integration of the different strands, and to strengthen the links with the industry's own representative structures.

And that brings me back to the crucial role of the Construction Industry Board. It is at periods of rapid change that the task of bringing together, discussing and forming a common view is both most difficult and most necessary. That is what Sir Michael Latham recognised when proposing the establishment of the CIB and that is what the CIB has sought to achieve. I pay warm tribute to the wise and skilful manner in which Tony Jackson has steered the CIB through these challenging past two years; to his positive response to the Movement for Innovation, as well as to his careful, quiet, insistence that the CIB and the umbrellas must be fully involved and integrated. I know that his task has not been easy. We all owe him a great debt for his sterling efforts. In his place, we are delighted to welcome Chris Vickers who has already made a number of important contributions to the work of the CIB.

There is no doubt that the industry will continue to need a strong central representative organisation. But the rapid changes we are going through may well require the CIB to be adapted, developed and strengthened ... We need a clearer vision of the way forward and of the institutional structure that will enable us to sustain the Movement for Innovation, to integrate the process of change into and throughout the fabric of the industry, and to create an atmosphere and approach that will enable the industry to ride the tide of change successfully.

At the same time, outgoing CIB chairman, Tony Jackson, announced the major review of its own role and responsibility:

> The message is simple. We know there are a number of initiatives being presented within the construction industry at the moment. There is a need to rationalise and bring more of the work together. We recognise that substantial progress has now been made in implementing the Latham recommendations, and that the construction scene has now changed considerably. The Board has been instrumental in delivering several new initiatives to the industry including National Construction Week, the Considerate Constructors Scheme and the Construction Best Practice Programme. Others too, including the Movement for Innovation and CBI's Fit for the Future campaign, are also contributing to the whole agenda of change in the industry. We believe that it is an appropriate time for the CIB to take stock of these changing circumstances and review how our work relates to these other bodies.

On 19 July the Deputy Prime Minister and Secretary of State for Environment Transport and the Regions, John Prescott, arranged to address the first Movement for Innovation (M4I) conference in Birmingham. In the event, because of other calls on the DPM's time, his address was delivered on his behalf by the Construction Minister, Nick Raynsford. He praised the progress being made by those involved in the Movement for Innovation, saying:

> It is just a year since the Deputy Prime Minister and Sir John Egan published Rethinking Construction and it is only eight months since the Movement for Innovation launch last November. Within a month the M4I board was established and by Christmas it was fully operational and had carried out a preliminary assessment of the first round of demonstration projects. By Easter the total value of demonstration projects was not far short of £3 billion – six times the target figure.
>
> In May we announced a set of key performance indicators enabling the industry to measure and compare its performance on a standardised basis for the first time. By the end of June each

of the demonstration projects had presented their innovations to the M4I board. By any standard this is a remarkable achievement in a very short time and I congratulate all those involved.

He went on to call for a much higher profile for sustainability in construction. He said that those at the leading edge of construction were already showing impressive results but that there was a need to keep up the pressure to convince a wider spectrum of the construction industry that sustainable construction reduces waste and costs, improves efficiency and increases profit margins.

The Movement for Innovation is issuing a call today for more demonstration projects showing innovation in sustainable construction. My challenge is to use these projects to prove to the wider industry, to clients, to the public, to the world market place, that British construction can indeed lead the world in innovative, effective approaches, and that sustainable construction is not only good for the environment, but also very good business.

In addition to this, however, the Minister on behalf of the DPM challenged the industry's clients to work together to give a more effective lead in dealing with the industry. He said:

Over the next year I would like to see public sector clients in both central and local government; and the other clients that are financed from public funds, especially in housing, education, health and transport, and private sector clients in the Construction Clients Forum and the Construction Round Table, all come together more effectively to drive forward change in construction.

I challenge the whole client community, both public and private, to come together and draw up a new clients' charter. A charter that sets out the minimum standards they expect in their construction procurement today, their aspirations for the future, and a programme of steadily more demanding targets that will drive standards up year by year. By this time next year I would like to see signed up and committed to that charter, enough clients to represent at least half of total UK domestic expenditure on construction.

This concept of a Construction Clients' Charter was initially envisaged, certainly in the mind of the Construction Directorate of DETR which provided the briefing for the DPM's M4I conference speech, as covering:

- measurement of project performance, including each M4I key performance indicator;
- measurement of client satisfaction with the supply side's performance and with the product, and measurement of the supply side's assessment of the client's performance;
- standards for, and monitoring of, health, safety, equal opportunities, site facilities, and the whole 'respect for people' agenda;
- standards for training;
- standards for, and measurement of, sustainability and elimination of waste, both in the construction process and in the product.

The new chairman of the client body, the Construction Clients' Forum (CCF), Terry Rochester, immediately, and without formally putting the matter to the Forum's members, publicly notified the clients' acceptance of the challenge, relying on his view of the likely constructive reaction to such a proposal by the majority of CCF members. Indeed, public sector clients under the leadership of the Government Construction Clients' Panel within the Central Unit on Procurement had been toying with the idea of some mandatory form of commitment to improvement in public sector procurement practice for some time. Terry Rochester, drawing on his previous experience of such processes within the Highways Agency, very quickly started work himself on designing and developing the charter concept.

The importance that the Government attached to this initiative was emphasised in a letter that John Hobson, director of DETR's Construction Directorate, sent to Terry Rochester immediately after the announcement at the Birmingham conference. John Hobson identified the key players in the exercise, in addition to himself from DETR, as Terry Rochester in his capacity as chairman of the CCF, John Kerman, the newly appointed quality director of the Highways Agency and chairman of the Construction Round Table, and Mike Burt, leading the Government Construction Client Panel's improvement programme within the CUP. Surprisingly no specific suggestions were made about private sector client participation; presumably the assumption was made that the Construction Round Table would provide such input. Whatever the real reasons were, however, this apparent failure early on to secure comprehensive private sector commitment to the development of the charter proved an unhappy portent in the longer term.

Under the shadow of these various developments, and with the evidence of the Government's apparent unease about the real influence of the CIB on the movement for improvement in the industry, the CIB's executive committee met on 22 July under its new chairman, Chris Vickers. The main business of the committee at that meeting was to agree the framework for the review of the role and function of the CIB which had been announced by the out-going CIB chairman, Tony Jackson. The review was to be headed by the new chairman, Chris Vickers. He confirmed that he saw the review of the CIB as a fundamental reappraisal of the structure and role of the CIB, not merely an examination of its existing agenda. To that end, he confirmed that the review would address the following fundamental questions:

1   What has the CIB achieved in the four years since it was formed?
2   Is there a continuing need for the CIB at all?

If there is a need,

3   What should its role and responsibilities include?
4   How should its membership be comprised?
5   What should its relationship be with other agenda and organisations, especially the Movement for Innovation?
6   How should it be resourced and funded?

He reported that discussions had already begun with the Government and the industry umbrella bodies, which currently comprised CIB membership. Other interested parties were also being consulted and a small steering group had been formed to distil and shape the ideas. The review was expected to take approximately ten months, with the final recommendations being presented to the current Board in April 2000.

Meanwhile, preparations proceeded within the client body, the CCF, to implement the recommendations of the review of the Forum's effectiveness undertaken during the previous year by its deputy chairman and CBI representative, Ian Reeves (see Chapter 7). While the CCF had been identifying and launching its 'Client's Pact with the Industry, Constructing Improvement', the Construction Round Table (CRT), with its considerable private sector membership, had issued its own Agenda for Change, much of which anticipated almost verbatim the aspirations in *Constructing Improvement*. The CRT's Agenda for Change was intended to be a statement of the very high standards sought by its members, focusing on designing the facilities they wanted as clients, the trading environment and the delivery process (see Figure 9.3).

The DPM's challenge to the clients, and the pressure that was being applied by John Hobson for some tangible evidence to emerge of major change in the industry's structure which the Government could publicly indicate, gave added impetus to the CCF to make rapid progress with the ideas put forward by Ian Reeves, and accepted 'nem. con' by the CCF's remaining member organisations. (Attendance at CCF meetings had begun to dwindle again after the temporary resurgence during the drafting and launch of 'The Pact with the Industry'.) Efforts were made by Martin Reynolds, chairman at the time of the CRT, and the CCF's chairman to effect a merger between the secretariats of the two client organisations, with

---

Designing our facilities
- Understanding and measuring our customers' needs;
- using design techniques and processes that address customer needs;
- identifying and managing risks;
- achieving predicted design performance;
- maximising benefits from standardisation;
- measuring customer satisfaction;
- continuous and sustained improvement.

The trading environment
- Objective, value-based sourcing throughout the supply chain;
- collaborating with suppliers;
- Improving our relationships with our suppliers and clients;
- establishing a no surprise culture;
- paying of suppliers;
- measuring performance in our supply chains.

The delivery process
- Employing efficient construction processes;
- creating integrated project teams;
- reducing waste and delivering with zero defects;
- moving from construction to production;
- Improving the quality and relevance of skills;
- measuring and publishing our performance.

Figure 9.3 The CRT's Agenda for Change.

the intention of reducing overheads and mitigating the worst effects on the CCF of its chronic under-resourcing. These efforts did not enjoy universal support, indeed there was a growing lack of mutual confidence, and they soon proved abortive. Under the CRT's new chairman, Peter Roberts from Transco, however, and in the light of the Reeves' recommendations, a joint CCF/CRT workshop was held, at which the CRT members publicly expressed their determination to establish and maintain a single client voice in the industry, and to 'pass the baton', in the CRT's words, to the envisaged new client body, the Confederation of Construction Clients (CCC). At this 'Baton Passing' workshop, representatives of the CCF and the CRT agreed that the proposed establishment of the new CCC as proposed in the Reeves review of CCF could open a new phase for clients in their relationship with the rest of the industry. Clients could achieve this by insisting that the new CCC should be driven by actual purchasing clients rather than by umbrella representative bodies, that it should focus on improving the industry from the clients' perspectives, that cultural changes reflecting the parallel efforts made by other organisations concerned with performance improvement, such as M4I, could be encouraged by clients adopting the Clients' Charter, and that owner/occupier clients, as well as clients buying construction to sell on, could achieve a higher profile in relation to the supply side by engaging with the proposed new CCC. This looked to be an encouraging position from the point of view of some of those in the CCF, the DETR, and other parts of the Government machine who were beginning to become disillusioned with the prospect of ever achieving a truly comprehensive and influential client body, and the CCF and its members, with overtly expressed support from the CIB, invested major resources in organising a major launch of the new body and the charter.

In the CCF, however, there was some sharp opposition to the CRT's bilateral (client/contractor) approach to procurement pushed particularly by prominent CRT members such as Railtrack, Highways Agency and BAA, as compared to the teamwork approach now being more widely used in building procurement. This difference in approach was a reflection of the Latham/Egan difference and was to torpedo the CIB.

The launch of CCC and the charter took place at the Royal Institution on 7 December 2000. Much of the organisation of the launch had been undertaken personally by the CCF Chairman, Terry Rochester, who had, with the agreement of some, but not all, members of the then executive committee of the client body, undertaken the duties of acting paid chief executive of the CCF. Considerable efforts were made to obtain formal support for the new CCC from all the major sectors of the industry, as well as from central and local government. Speakers at the launch were Nick Raynsford MP, who had by this time become Minister for Housing and Planning with Beverley Hughes MP supporting him on local government affairs and construction, Digby Jones, the director-general of the CBI, Peter Gershon, recently appointed as chief executive of the Office of Government Commerce which had developed as a separate government department from the old Central Unit on Procurement, Richard Arthur, leader of the London Borough of Camden representing the Local Government Association, Tony Giddings, chairman of Argent plc representing the private client sector, Alan Crane, chairman of M4I and Michael Dickson, partner in Buro Happold and chairman of the construction professionals umbrella body, the Construction Industry Council.

At the same time as the launch of the new CCC, the new organisation also launched the Clients' Charter, development of which had been promised by the CCF in response to the challenge issued by the Deputy Prime Minister the previous year. Here again much of the detailed work in designing the charter, which sought to embrace within it a codified system of procurement best practice, was undertaken by Terry Rochester personally, building on the experience of the Highways Agency. The charter had always been seen, from its inception, as being inclusive of the supply side of the industry, as well as clients. The objective was to give tangible form to the accepted concept of partnership between client and supplier, and four pan-industry workshops had been held during the year preceding the launch with the aim of achieving this commitment from the supply side. Although through the Construction Industry Board the supply side umbrella organisations had given firm indications of their willingness and enthusiasm to join in developing the charter, this potential consolidation of the principle of collaboration was not properly exploited by the client body. This failure properly to involve the supply side proved an inhibition on acceptance of the charter by a wide variety of clients, and necessitated corrective action later.

By the time of the launch in December 2000, Mike Roberts, a member of the CCF executive committee and about to retire as group technical director of BAA plc, had accepted to become the new Confederation's first chairman. At the launch he was able to announce that full membership of the Confederation included:

BAA plc
Defence Estates
Highways Agency
Tower Hamlets LBC
McDonald's Restaurants
NHS Estates
Office of Government Commerce
Railtrack plc
St Helens MBC
University of Surrey

and that the Confederation would concentrate on:

training:	achieving a fully badged workforce of compliant operatives; managing training for the supply side; developing client skills;
measurement:	having a preferred, consistent and continuous system for measuring the effectiveness of projects;
briefing:	improving the whole process of client–supplier briefing;
integrated supply chain:	guidance to clients on creating an integrated supply chain and construction teams;
selection criteria:	developing consistent criteria for appointing suppliers, e.g. safety, health, equal opportunities, environmental sustainability, integrated teams.

As evidence of Confederation members' commitment to advancing these objectives, membership was made conditional on the particular organisation agreeing to sign up to the Clients' Charter.

The launch of the CCC and the Clients' Charter was widely welcomed. The *Financial Times* (7 December 2000) saw it as a 'new alliance to boost construction standards' and expressly reported Mike Roberts' public denial that 'clients had got together to gang up on contractors'. In the technical press, *Contract Journal* saw it as potentially marking a 'real milestone for the industry'. 'If,' the *Journal* said, 'the CCC proves to be more than just a talking shop for major construction clients and is instead a major driver of change, then the industry really does have a chance of ending its adversarial and inefficient culture.' *Construction News* saw it as the 'new organisation designed to provide a single voice for construction clients'.

The euphoria engendered by the launch of the CCC and the charter encouraged the new Confederation to seek a high-profile, dynamic chief executive to direct its affairs and undertake a major recruitment campaign, and to seek a commercial partner to operate and develop the charter on its behalf.

This newfound confidence was, however, to be severely dented by adverse developments. The first was the unexpected decision of the British Property Federation (BPF), the trade body representing the important property development client sector, to have nothing to do with the new client movement. The BPF set out its reasons for this decision in a letter to DETR's construction director, John Hobson. This is worth quoting in full, since it incorporates a number of potential weak points which adversely affected all the efforts made since Sir Michael Latham reported to achieve a genuinely representative client movement (see Figure 9.4).

Geoff Wright, a former CCF chairman and director UK Property at Hammerson's remains of the opinion (letter 28 April 2004) that it should have been of no surprise that the BPF declined the invitation to join the new CCC. In his view they had not been consulted about the line adopted by the CCF in relation to the future of the Construction Industry Board, and had been offered membership terms for the new CCC based on the estimated turnover value of the Federation's member firms, resulting in a proposed level of corporate subscription unacceptable to a trade association.

The withdrawal of the BPF was followed, for similar reasons, by a decision by the Association of Directors of University Estates not to support CCC following its unsuccessful and largely ignored efforts to get the CCC chairman to accept that a teamwork approach was in many cases more appropriate than total emphasis on client/main contractor bilateralism.

Another, more serious, setback to the success of the new dispensation related to a sudden change in input by the CCF, the predecessor of the new CCC, to the review of the Construction Industry Board (CIB) which had been initiated by the Board's chairman, Chris Vickers, and which had been underway during the 12 months or so leading up to the CCC's launch.

Chris Vickers had initiated the review of the CIB immediately on taking up the CIB chairmanship, with the support of all the major players in the industry who had, in one way or another, been associated with the work of the CIB or in the other pan-industry activities

John Hobson
Director of Construction
Zone 3G6
DETR
Eland House
Bressenden Place
London SW1E 5DU

24 November 2000

BPF
BRITISH PROPERTY
FEDERATION

1 Warwick Row, 7th Floor
London SW1E 5ER
Telephone (020) 7828 0111
Facsimile (020) 7834 3442
e-mail: info@bpf.org.uk
website: www.bpf.org.uk

Dear John

Construction Clients Confederation

I wish to inform you that the BPF has decided not to join the new Construction Clients Confederation.

This is not a decision the BPF has taken lightly. The property industry is one of the most important clients of the construction industry and it has a vested interest in seeing the Government's aim to improve the efficiency of the construction process succeed. The industry's commitment to that end is demonstrated by the fact that the BPF, together with CIPS, was instrumental in setting up the CCF in 1994 and has provided a substantial part of its administrative and policy leadership since then.

The BPF was broadly satisfied with the aims and structure of the CCF and believe benefit was derived from membership. However, the CCC appears to be taking a very different route and for the reasons set out below, we do not believe it will best serve our interests or lead to tangible benefits to our sector.

In summary, the aspects of the CCC which deter us from joining are:-

- The membership is not representative enough of the private building sector (nor ever likely to be) being dominated by:-
  a) public sector bodies;
  b) civil engineering and other non-building parts of the private sector;
  c) large corporates.
- Partnering is seen as the preferred means of procurement and standard contract forms, including JCT, are not seen as a core CCC interest.
- Participation in the CIB appears not to be regarded as a core CCC interest.
- The views of trade associations appear not to be valued compared with those of single, albeit large, individual contracting clients.
- Membership of the clients charter is a pre-requisite of membership.

The above are core points of principle which remain valid regardless of any other consideration. However, I should point out that the subscription being sought from the BPF for collective membership is unacceptable also.

It is clear from a meeting of BPF members held to discuss membership of the CCC, that, because of these issues, it is very unlikely any BPF constructing members will join the CCC.

The result, therefore, is that the CCC will not be mandated to speak on behalf of the property industry, a major source of construction clients. In view of this, I request that, as the trade association of the property industry, the BPF is consulted directly by the DETR on all construction issues from now on. This was the arrangement which existed before the formation of the CCF when both sides gained clear benefit and I have no doubt that this will be achieved again.

Yours sincerely

WILLIAM MCKEE
DIRECTOR GENERAL

Figure 9.4 Letter to John Hobson, DETR, from the BPF's Director General, William McKee, setting out its reasons for not joining the CCC.

set up to work in parallel with it. His decision reflected the unease which had become increasingly apparent among the Construction Confederation members about the real level of support they felt for the CIB, as well as that of the new Confederation of Construction Clients which was finding it difficult to bring the influential private sector client organisations within its membership. The reluctance of these clients to become members of the client representative body again resulted from their suspicion that the main contractor sector of the supply side generally doubted whether the clients, particularly those in the private sector, were truly determined to jettison the 'take the lowest tender price' approach to the award of work. This circular growth of mutual distrust risked a return to the poor relationships within the industry which had led to the need originally for a wholly new approach to construction business. Influential also in the industry's thinking at the time was the suspicion that the Department's construction director, John Hobson, and the ministers that he was advising, were losing patience with what appeared to be slow progress in achieving and implementing the reforms that had been identified, and were only supporting the activities of the newly-formed M4I and the Egan implementation group.

Recognising the importance of regaining trust and momentum from the main players, Chris Vickers established a steering group for his review, composed of the chairmen and deputy chairmen of the CIB umbrella bodies; Robin Nicholson for the CIC, Alan Crane for the contractors, John Harrower for the CLG, David Adamson for the CCF, Roy Harrison for the Construction Products Association (CPA) (the component and material suppliers) with John Hobson from the Department covering the Government's interest in ensuring that overall sight was not lost of the need ultimately to establish a body for the industry that would prove genuinely strategic. These accredited representatives of the constituent bodies were supported by their organisation's chief officers, and the review itself was serviced by the Board's chief executive, Don Ward. It was agreed that only the above would be taken to represent their constituencies.

Chris Vickers was sensitive to the prevailing atmosphere among the different industry sectors and, having taken soundings from a wide circle of consultees, announced early on that the review would genuinely be 'root and branch', including deciding whether a pan-industry strategic body was needed at all. By February 2000 he had issued a comprehensive consultation paper in which what appeared at the time to be the universally accepted opinion that a strategic body was indeed necessary was reflected in his concept of a 'Newbody' successor to the Board.

In his consultation paper Chris Vickers said:

> Every person or organisation to whom I have spoken during the course of this Review has told me that a strategic forum of the industry bodies, clients and Government is required for policy and standards-setting. We need a pan-industry forum of the representative bodies of the industry at which senior representatives can discuss and agree issues of common interest where they can have an impact.
>
> This would be a body seen to be 'owned' by the industry and able to provide strategic leadership and influence. The debate of the strategic issues that affect all in the industry is best conducted by senior practitioners of the industry and, if the forum is to attract the contribution of the people it requires, they in turn need to be persuaded to give of their time and that it is in their commercial interest and that of the industry's that they do so.
>
> The vision of this body would be an efficient and successful industry which fully meets its clients' needs and expectations, and in which clients and Government act to make that possible. Its mission would be to bring about through pan-industry strategic leadership an efficient and successful construction industry that fully meets clients' needs and expectations. To achieve this, its objective would be to identify and develop policies of strategic importance to the whole industry which will lead to improvement in the quality of resource, the quality of product, the efficiency of operations, and profitability in the industry, whilst providing its clients with better value for money.

He also drew attention to the expansion in the number of industry initiatives (there was much criticism around at that time of 'initiative fatigue'), and the need to rationalise these:

> The other strong influence on my thinking is the large number of other organisations which exist to deal with a wide range of construction-related activities, some pre-dating the CIB and others independent of the CIB. The phrase 'initiative fatigue' has come into common use – sometimes as an excuse to justify inaction, but nevertheless a problem sufficient to merit specific attention as part of the Review.

Those which I judged most important to the Review were the Construction Best Practice Programme and the Movement for Innovation – the one set up very much at the CIB's recommendation, the other without any CIB involvement, in itself arguably an indictment of the CIB's failure to provide the right leadership. But altogether I took account of more than fifty 'initiatives'. Given the CIB's desire to lead or 'facilitate' by consensus, this has given rise to a loose and potentially uncoordinated framework when what is required is greater cohesion and influence to avoid duplication of effort and to produce greater efficiency and more rapid progress in dealing with the challenges facing the industry.

Chris Vickers titled his consultation paper *'Newbody': A Root and Branch Review of the CIB*. In it he proposed a new pan-industry organisation to replace the CIB. This 'Newbody' would:

• provide a single representative strategic forum of the industry, its clients and Government;
• develop and drive a pan-industry strategy of improvement based on innovation, demonstration, communication, implementation and monitoring;
• add value by addressing only pan-industry matters which could not be dealt with at a subsidiary level; and
• engage with and influence the activities of the wide range of industry-related 'satellite interests'.

It would concentrate on four core strategic interests, formulating annual business plans to facilitate improvement in:

• sustainable performance and best practice;
• image of the industry;
• innovation and research strategy;
• facilitating Government sponsorship of the industry.

Bearing in mind the industry's wish for a rationalisation of the 'multiplicity' of initiatives, which was perceived as slowing down progress through confusion or duplication, and 'turning off' some volunteers, he proposed that Newbody should use its strategic influence to facilitate a more coherent framework for industry improvement with 'satellite interests' for whom the avoidance of duplication and wasted effort was a priority, and with the vast majority in the industry who remained completely uninvolved and who just wanted to make sense of the apparent plethora of initiatives. At that time there was much unsettling speculation as to how long M4I would continue. In this connection Chris Vickers' report anticipated that M4I would not continue beyond its originally envisaged two-year life and would provide during that time the 'test bed' for new ideas relating to radical change. The report envisaged 'Newbody' directing, and setting priorities for, the work of the Construction Best Practice Programme (CBPP) in disseminating 'best practice': the vision was to consolidate and focus the range of overlapping reform initiatives.

The consultation paper proposed that membership of the new body should be made up of the five umbrella bodies representing the various sectors of the supply and demand sides of the construction industry:

- the Construction Industry Council (professional institutes, consultants' associations, research organisations and standards-setting bodies);
- the Construction Industry Employers' Council (lead contractors);
- the Construction Liaison Group (specialist contractors);
- the Construction Products Association (materials and product suppliers);
- the Construction Clients' Forum (the clients of the industry's products and services from the public and private sectors);
- the Government (as sponsor for the industry), led by the Department of Environment, Transport and the Regions.

The Government's involvement was seen as involving its three roles, as client, regulator, and sponsor. Chris Vickers' proposals were that:

its client interests would continue to be catered for within the CCF/CCC;
its regulatory interests would be included by invitation of the appropriate body (e.g. DETR, HSE, DfEE, etc.) to relevant policy discussions;
the Government's sponsorship role – i.e. support for improved competitiveness of firms in the industry – is currently provided by the DETR. This is the role which, for most other industries, is provided by the Department of Trade and Industry. It was envisaged that on occasions the Government's 'sponsor-

ship' role would require involvement of the sponsor department (DETR) in dialogues with other Government departments (e.g. HM Treasury, Inland Revenue, DfEE, DTI) to better understand the industry position on particular issues.

Chris Vickers' proposals for the structure of the new pan-industry body envisaged a management board with each participating organisation, including Government, having two seats each, and a chairman's advisory group incorporating one representative each from participating organisations, and representatives as appropriate from time to time from other relevant interests, on invitation from the chairman and other members of the advisory group.

As regards resources, the report recommended a higher level of strategic analysis and support than had been available to the CIB. Initial estimates by Chris Vickers put the funding required at about £250k for the core operation, much the same as available to the CIB. He recommended that initially 50 per cent of the funding required should be provided by Government, with this proportion reducing over three years until it was entirely covered by client and supply side sharing the required amount equally.

Chris Vickers' consultation paper was circulated in February 2000. Although responses were requested by early April, organisations asked for more time and it was not until May that reactions were collated and debated within the CIB executive committee. A tight deadline had been imposed because of the chairman's declared intention, issued following strong direction from DETR, of announcing the outcome at the CIB Annual Consultative Conference at the end of June.

The general reaction to the consultation paper was not encouraging. All the main participants recognised the skill with which Chris Vickers had tried to cover the many disparate issues, but the very fact that the proposals required them to face up to the many unresolved, and in some cases unarticulated, adversarial interests, brought out into the open the underlying vulnerability of the relationships between the various sectors.

Concern crystallised around five main issues:

1   the view of the main contractors (represented by the CIEC, soon to become the CC) and the clients (represented by the CCF in process of evolving into the CCC) that Newbody was merely CIB Mark 2, and did not recognise CIEC's and CCC's conclusion that CIB had really failed to deliver, and that an entirely new approach was needed;

2   the strongly held view of the private sector client members of the Construction Round Table that what was needed was a framework within which clients and suppliers could meet face to face, without designers and specialist contractors, and in which Michael Latham and John Egan's vision of an industry led strategically by its customers could be realised, and the equally strongly held counter-view of designers, suppliers and some clients;

3   the main contractors' view that, because their representative body, the CC, covered so many different types of contractors, they should have more seats on the management board than the other supply side sectors (professionals, sub-contractor specialists, component and material suppliers);

4   refusal by both the main contractors and the clients to commit themselves to finding the necessary financial contributions;

5   the agreed view of contractors and clients that Government should not be formally a member of the new organisation, but should attend by invitation of the members.

Given this situation, Chris Vickers immediately recognised that if progress was to be made it was vital that the clients were supportive. As noted earlier, the CCF had asked David Adamson to be responsible for the clients' collective input to the review and its implementation. An initial client response to the consultation paper calling for a 'collegiate' structure in which clients would speak collectively and, on the basis of equality, direct to the supply side, was prepared and sent to Chris Vickers, and David Adamson met the CIB executive committee on 19 April.

At the meeting, Chris Vickers and the chairmen of the other umbrellas (with perhaps some slight reservation on the part of the Construction Confederation) recognised client dissatisfaction with the current way of working of the CIB and the difficulties this posed for CCF/CCC in bringing on board the top practising clients they wanted to involve in the future industry strategic body. Other umbrellas noted similar problems in getting attendance by the best people.

In his report to the CCF/CCC Executive David Adamson said:

> Chris Vickers will, in the end, write his report as he was requested, to Government; we must do our very best to persuade him to our views.
>
> All support the main thrusts of the aims and requirements of Newbody, except for CIEC alone calling for M4I/CBPP to continue independently: we should continue to oppose that.

We propose a bilateral collegiate structure: this has so far not attracted any support but an offer of larger representation of clients to balance the supply side at the table is proposed as a means of ensuring a 'hard edged' client/supplier dialogue between key practising professionals.

There are shades of opinion but general support that Government attendance should be by relevant stakeholders in 'observer' status, contributing to but dominating neither agenda nor meetings.

I see the alternatives as a constructive and business-like Newbody on the one hand, and continued fragmentation and duplication on the other if we fail.

Subsequently, David Adamson informed the CCF executive that he considered it both impractical and illogical for the clients to demand a 'one to one' forum with the suppliers, given the variety of sectoral interests that 'Newbody' was intended to bring together. He wrote:

What is called for is:

a An organisation that will have strategic and deep discussion of key issues as these arise so as to set policy and improvement across the industry, including the plethora of initiatives. It is a 'given' that discussions must be demonstrably effective.

b Clients and so-called 'supply side' (definitions in much of the industry are not simple), each with roughly equal membership, meet on an equal basis with contributions from those with useful input. Such client membership would allow public and private, large and medium clients.

The above are achievable.

It is the experience of most practising clients that useful and deep debate on key, difficult issues cannot be achieved unless significant practising members of the industry can contribute when they have something useful to say (keeping quiet when not). On the other hand, 'across the table' dialogue between just two 'voices' – one 'voice' representing the wide spectrum of clients and just one 'voice' representing the even wider spectrum of designers, constructors, pfi clients, suppliers, etc. – does not lead to deep, rich, 'owned' and widely-informed discussion. The evidence and reasons for that are clear; some were set out in Michael Latham's report which took the industry beyond the 'two voices across the table' approach. On grounds of logic and practicability I cannot support the 'two voices across the table' proposal.

At a key CCF meeting, there was a balance of support for this view and that was due to be delivered at the final Newbody review meeting in May 2000. CIC and the specialist contractors made clear their support for this stance. It appeared that this was where the matter would rest, but events took an unexpected turn. David Adamson was unable to attend the Newbody review meeting (he was moving house) and it was assumed that the written position of the CCF would be taken since he was the delegated representative of CCF. However, Terry Rochester, without consultation with others, turned up and was allowed to speak to a different brief, that of the bilateral (one client voice to one supply-side voice) favoured by the CRT.[1] This called for:

1 a bilateral supplier/client forum with equal status for both;
2 a collaborative framework for all of the industry to work together, with the minimum possible administrative support structure;
3 the clients with their own secretariat and chairman meeting to discuss productively strategic issues with suppliers who, once it can be achieved, would speak cohesively having addressed supply chain issues in their own routine discussions;
4 cost beyond the minimum support structure (as noted above) to be met collectively where appropriate by a party achieving particular benefit;
5 other 'influences' on the industry, including Government, to be invited to input as and where appropriate; on most issues there would be input from some part of Government.

With the short-time scale until the CIB consultative conference rapidly shrinking, Chris Vickers did what he could to meet the concerns expressed, but he wrote to the CCF chairman, Terry Rochester, in response to the CCF submission. In his letter he said:

With the clients on board the new CIB would be a stronger organisation than without, but if that is indeed your position then I shall have to proceed accordingly when the time comes for the implementation of my proposals, which I am confident will be accepted by the other interested parties.

This indication that the review of the CIB was prepared to at least consider a future strategic body for the industry which would not include client participation started many alarm bells ringing, and indeed did not augur well for the future. Terry Rochester, chairman of the CCF and its acting chief executive, sent a placatory letter to the CIB chairman which, while denying that CCF members from both public and private sectors would refuse to support the proposed Newbody, still held out for the establishment of a 'supplier client forum with equal status for both ... seeking wherever possible to define jointly what needs to be done ... to implement this in a collaborative way, utilising the industry's resources, including those of the clients, and with the minimum possible administrative resources'.

With the deadline for presentation of his review at the CIB consultative conference now very close, Chris Vickers sought to interpret this somewhat ambiguous assurance by the clients' body in the best possible light. In his reply he thanked the CCF chairman for 'clearing up my misunderstanding', but he emphasised that 'my round table concept for the new body is, I believe, a fundamental requirement if it is to succeed and client participation in this form is essential'.

Chris Vickers now finished his root and branch review for presentation as 'A Framework for a Strategic Group for the Construction Industry'. Much of the content built on the proposals in his consultation paper, but to meet particularly the clients' stated wishes, the clients were given six seats on the board of management, with the four supply side bodies having two seats each. The Government industry sponsor was not formally a member, but its function as client and regulator was included through the proposed client and supply side representation. The earlier proposals for the Newbody to assume responsibility for, and direction of, the other Government-backed activities, particularly M4I and CBPP, were glossed over, although a specific bid was made for Newbody to take over CRISP (Construction Research and Innovation Strategy Panel).

Requirement for funding was put at £250k per annum, with the proposed participating bodies and Government adjusting their proportionate contributions over three years as shown in Table 9.1.

Where the new body identified a need for project-specific finance, the sponsoring Government department would give high priority to formal proposals supported by the new body for matched funding of up to 50 per cent of project costs.

The legal identity of the new body was proposed as a company limited by guarantee, with the directors comprising the chairman,

Table 9.1 Proposed funding for the new bodies

Organisation	2001 (%)	2002 (%)	2003 (%)
CCC	21	28	35
CIC, CIEC, CLG, CPA (each)	7.5	9.5	12
Government sponsor	49	34	17

chief executive/company secretary, and one director appointed by each member body. This was intended to provide the flexibility to employ staff or enter into contracts for the supply of other services.

As noted earlier, this new organisation was formally announced at the CIB consultative conference at the end of June 2000, and given a guarded welcome by the Government. Any hopes that the basic differences in approach that had become apparent during the consultation period from both the main contractors and the clients had been resolved were, however, soon to be dashed.

The CIB executive committee took on the task of developing the 'Framework for a Strategic Group for the Construction Industry', and for implementing the review's recommendations. Immediately it became apparent that the main contractors did not see themselves as committed to the continuation of the pan-industry representative body concept, despite having gone along with the earlier discussions. At an early stage they made clear that they would not contribute financially as proposed, since in their view it was unnecessary for a pan-industry representative body to have permanent staff, offices, or a programme of executive action. This, they felt, was better dealt with by bilateral action between the parties concerned. The main contractors would also not accept equal status in terms of seats on the Board with the other supply side sectors.

The action of the client body proved even more difficult to understand, and in the end lethal to the idea of a new powerful and comprehensive pan-industry body. Following the presentation and apparent acceptance of the CIB Chief Executive's review at the CIB consultative conference in June 2000, the CIB decided to establish a board of management to implement the proposals for the Newbody, to run in parallel with the CIB's executive committee, and take over from it when it proved appropriate to do so. Meanwhile, within the client's organisation, the CCF, action was being taken to establish the new Confederation of Construction Clients (CCC), due for official launch in December. A new chairman of CCF to oversee the transition to the new CCC was elected. This was Mike Roberts of BAA, and in the light of CCF's wish to press ahead with successfully expanding the client body the former chairman, Terry Rochester, was asked by him (despite misgivings from some of the remaining members of CCF) to undertake the duties of temporary chief executive of the organisation on a paid basis. It immediately became apparent that neither the new chairman, nor the new temporary chief executive, felt themselves bound by the CCF's previously expressed commitment, albeit with reservations, to supporting the Newbody. At the final meeting of the CIB executive, held in September 2000 formally to set up the Newbody's board of management and accept initial proposals for its strategic approach and medium-term programme of work, Terry Rochester explained that, while CCF, soon to become CCC, accepted the broad principles of 'Framework for a strategic group for the construction industry', they were unable to speak on behalf of the yet to be formally established CCC or to discuss any of the detail of that document.

This indication, relatively unexpected in view of what had been said earlier, of reservations on the part of the clients to a Newbody seems to have resulted from strong pressure by some major client members of CRT to return to a situation where they were able to meet main contractors face to face and exert pressure on them as a result of the purchasing power they were able to deploy. As a result, although the main contractors did not oppose the 'framework' proposals, this change of tack by the clients confirmed an inbuilt reluctance to commit themselves to the round-table concept put forward by Vickers. The other supplier umbrella bodies (CIC, CLG and CPA) continued to support the framework as proposed. They expressed grave concern about this declared client stance.

The first meeting of the Newbody board of management was held at the end of October, and the new chairman of CCF/CCC, Mike Roberts attended to report on the clients' position, but without any commitment to join the new organisation.

Asked about the CCC's attitude towards joining the 'new CIB', Mike Roberts explained that the CCC had not yet agreed its position. However, he acknowledged that it would be important for the CCC to find ways to have a dialogue with the CIB's umbrella bodies, and he wished to persuade the CCC to join in. He saw the offer of six seats as positive, as this would enable the CCC to bring an eclectic mix of clients from different sectors. Once the CCC had developed its proposed work plan, they would wish to share this with CIB member bodies.

Referring to funding, Mike Roberts explained that the CCC was in its infancy and faced a considerable task to recruit members and finance the data management organisation required for the Clients' Charter. The CCC was thus in no position to contribute funding to the 'new CIB' at this time. The CLG expressed concern at this position, and Mike Roberts agreed to take this back to the CCC. Mike Roberts also reported that the CCC would have a three-year life, and wondered whether the 'new CIB' should emulate this, rather than five years as proposed. The chairman acknowledged this position.

Mike Roberts stressed that, in any case, members of the CCC would seek direct dialogues with the supply side as well as through the 'new CIB'. Chris Vickers confirmed that this was precisely the same policy – and practice – adopted by all the supply side umbrella bodies.

By the time of the board of management's second meeting in December, the question of whether or not the clients would participate in the Newbody was still unresolved. The board of management, and DETR representing the Government's interest, agreed that a single client voice was desired, DETR and others stressing the need for a single strategic forum which included a demand voice, and that the supply side umbrella bodies did not wish to endanger this. It was agreed to ask the CCC to indicate by the end of February 2001 whether and how it would take up the six seats offered on the new board of management, so that final decisions on membership of the 'new CIB' could be taken by the end of March. It was rumoured that the Minister, Nick Raynsford, strongly expressed his disappointment to Mike Roberts at the client stance.

The 'new CIB' continued to work on the development of its strategy, and on implementing the programmes of work on sustainability, image of the industry and human resources, and facilitating Government sponsorship, as well as continuing close liaison with the M4I and Rethinking Construction movements, but it was plain by now that the idea of a long-term Newbody to succeed the CIB was doomed. This indeed proved to be the case, as Chapter 11 records.

Note

1 The news of Terry Rochester's attendance to and input at the meeting came in a series of worried telephone calls to David Adamson on his mobile as he was trying to unstick a bookcase from a bend in his new staircase.

10 The work of M4I and Rethinking Construction

The origins of Rethinking Construction; discussions within Government leading to the appointment of Sir John Egan; John Egan's brief; decisions on what industry interests and sectors should be involved; policy objectives of ministers, including Secretary of State John Prescott and Construction Minister Nick Raynsford; views of Construction Directorate of DETR; official views of collaboration with, and involvement of, CIB; emergence of M4I; aims and objectives; concept of demonstration projects; coverage of demonstration projects; development of key performance indicators; emergence of formal structure for M4I; availability of resources; Government and industry support; appointment of Alan Crane; current and future programme of work; description and assessment of achievements to date and implementation of Rethinking Construction principles. References to other sections.

Even if the inability of the Construction Industry Board to continue to inspire the support necessary to ensure its long-term position as a genuinely industry-wide strategic body proved disappointing, there can be no doubt that its work convinced the industry and Government that the approach adopted following Latham's recommendations was the right, indeed the only, way forward for the industry's success. This determination not to forfeit the advantages that were beginning to be recognised, particularly in relation to the adoption of partnering as the basis for the supplier/customer relationship, was manifested in the establishment of Rethinking Construction and the Movement for Innovation (M4I).

As described in Chapter 8, the ideas for an approach based on rethinking, and then re-ordering and implementing, the process of construction originated with the Labour Government, newly elected in 1997 after 18 years of being, if not in the political wilderness, certainly out of power on the national scene. Alan Crane, at the time deputy chairman, and on the point of becoming chairman, of the contractors' trade association, the Construction Confederation, and himself active in Labour Party politics at the local level, has recorded (interview 28 April 2004) the starting point adopted by the newly appointed Secretary of State for the Environment and Deputy Prime Minister, John Prescott. This was that the size, urgency and national importance of the new Government's programme of infrastructure improvements were such as to lead it to adopt a formal policy that the spending plans of the previous administration would not be altered for a period of two years, while the new Government worked out its national priorities in the light of developing economic circumstances. With this commitment to a very substantial programme of infrastructure investment, the Chief Secretary to the Treasury, Geoffrey Robinson, also made clear that before the programme could start, there had to be evidence of an appropriate level of competence from an industry that had, in the eyes of the new Government, a poor record of providing value for money; performance was defined as providing products and services on budget, to time and of appropriate quality to meet users' requirements.

The background to the Deputy Prime Minister's decision to appoint Sir John Egan to chair a Construction Task Force, mentioned in Chapter 8, is unclear. Alan Crane, the chief executive of the construction company Christiani and Neilson, and, as mentioned above, just about to take over as chairman of the Construction Confederation, is adamant that no previous discussion with the industry took place. He describes how, as chairman of the CC, he was about to open an 'invitation-only' conference of 350 construction industry firms' chief executives at the Dorchester Hotel, at which the DPM and Secretary of State for the Environment John Prescott was to deliver the key-note address. At 8 a.m. on the morning of the conference he, Sir Martin Laing who was also to speak at the conference, and Confederation deputy chairman, Paul Sheppard, were asked to meet the DPM at the Environment Department's offices, where they were told that he had appointed Sir John Egan to 'look at' construction and that this appointment, together with the names of 12 members of a Construction Task Force, would be announced at the conference.

The choice of Sir John Egan was a personal one by the DPM, who had worked with him previously and who had much respect for him as a successful 'hands-on' businessman, with a propensity for action and a dislike of formal structures and bureaucracy. Sir John was also a declared Labour Government supporter and influential in Labour Party policy making circles. As a major client of the industry, in his capacity as chief executive of BAA, he also had strong views about the industry's apparent tendency to ignore their customers' requirements, providing them with solutions based on the suppliers' rather than the clients' perceptions. The story is that Sir John, standing next to John Prescott at an industry function, had asked him what he intended to do about construction, eliciting the reply that if Sir John would undertake to shake the industry up, he would have the Government's whole-hearted backing. Other interpreters of this anecdotal history have pointed out that BAA were particularly keen to have an efficient construction industry, should they have a positive decision on the long drawn-out public enquiry on Heathrow Terminal 5, then with the Secretary of State.

Robin Nicholson (partner in Edward Cullinan Architects and one of the representatives of the Construction Industry Council on the CIB) has drawn attention to the 'crude but effective sticks and carrots' which Sir John presented to the supply side of the industry, particularly his strong indication that major clients would bring in contractors from other European countries if the supply side failed to deliver better client-focused products. Nicholson quotes the figure of £1bn as the annual costs of defects pertaining at the time.

Whatever the background to the appointment, its announcement at the conference caused consternation. After the DPM's departure, a 'vocal' (in Alan Crane's words) discussion ensued with some chief executives, including Sir Neville Simms, on behalf of the main contractors, expressing outrage, and asking who these members of the Task Force were to tell the industry how to do its work. As pointed out in Chapter 8, none of the 12 members of the Task Force had been previously involved with the activities of the CIB, or even more widely with the follow-up to the Latham recommendations. Although protests about the existing consultative machinery of the industry being left on the sidelines continued, there was recognition also that if the Rethinking Construction movement was to gain credibility it was necessary to obtain the commitment of acknowledged decision-makers in the industry who were themselves actively engaged in contracting. Some of these, who ultimately accepted invitations to join the task force, had been sounded out, prior to the election, at a private dinner party attended by John Prescott, Gordon Brown and Peter Mandelson and addressed by John Egan. Some commentators drew a connection between the political and industry tensions that flowed from this meeting (and perhaps from other meetings of such nature). Despite strong claims to the contrary, it might be expected that political decisions and positions would be taken on this major industry prior to the election.

John Egan and the Construction Task Force pushed ahead with preparation of the task force's report, with the full support of the Department. Despite a widely held view to the contrary, Alan Crane categorically refutes suggestions that preparation and writing of the report was undertaken by Simon Murray, a colleague of John Egan at BAA. Drafting of the report, according to Alan Crane, was undertaken by Jeff Channing, a senior civil servant in DOE, working to DOE's construction director, John Hobson. This is not to imply that the report was prepared as a quasi-Governmental project. On the contrary, John Egan himself, and members of his team, were closely involved personally in its drafting, and approved and amended particular sections as they emerged. As mentioned in Chapter 8, in an attempt to keep the CIB involved in this development of policy, the CIB chairman, Tony Jackson, established, with the agreement of John Hobson's Construction Directorate in DOE, a 'shadow' Egan group within the CIB, which was intended to ensure that work previously undertaken within the Board was appropriately covered by the task force, duplication minimised, and the momentum established by the Board maintained.

The task force, and the CIB's Egan 'shadow group', had a first sight of the draft report in May 1998. It immediately became apparent that there were significant shortcomings in coverage, in that the report made no mention of the important areas of housing and sustainability. This was particularly unfortunate in that, as pointed out by Alan Crane personally to John Hobson, part of the Secretary of State's declared responsibilities related to the need to provide suitable housing projects to meet social needs, at a time when the preoccupation with private house ownership was beginning to lead to large increases in house prices. Accordingly, the recently established Housing Forum was quickly involved in the task force's deliberations, and a section on the implications of the task force's findings for private and public housing added to the draft report. The impression remains, however, that this contribution was very much in the nature of a late addendum to a report, the general lines of which had already been decided and which the writers of the report were reluct-

ant to change in principle. Again, it was only as a result of intervention by the Government's adviser on green issues, Dr Jonathan
Poritt, that reference was eventually included to sustainability.
When it emerged, the task force's report, *Rethinking
Construction*, was given a rather low-key launch at a press conference which Prescott, Raynsford and Egan held jointly. The task
force's recommendations (see Annex to Chapter 8) concentrated on
three areas: identification and dissemination of lessons to be learnt
from a series of demonstration projects; establishment of a dynamic
movement for change (the Movement for Innovation (M4I)); and the
accessibility, through a best practice programme and a knowledge
centre, to excellence in practice.

John Hobson's concern, as DOE's construction director, was with
how the report's conclusions and recommendations could be speed-
ily and effectively implemented. In an understandable attempt to
ensure that the CIB and the work that it had already undertaken was
integrated with that of the task force, the CIB's chairman, Tony
Jackson, recommended that the Construction Industry Board should
be officially appointed by Government as the body charged with
implementing *Rethinking Construction*. It appears that this recommendation by the CIB was never officially rejected, either by John
Hobson or by Sir John Egan and his task force, but merely left in the
air with no indication of whether or not it was regarded favourably.
It was, to all interests and purposes, effectively ignored.

What is certain, however, and what was made clear at the time, is
that Sir John Egan and the Department did not regard the *Rethinking Construction* report as an attempt to achieve improvements in
the existing industry, but rather as an accepted conclusion that it was
necessary for the industry to act in an entirely different way to that
which had been the tradition hitherto. It was for this reason, according to Alan Crane, that it was felt that the existing industry representational bodies, committed as they had to be to the protection and
development of their own members' interests, were unsuitable vehi-
cles for taking the issues forward. The task force, and by implication
the Department, thought that it would be a fundamental mistake to
refer the specific Rethinking Constructing proposals to established
working groups of the CIB, some of which were already working on
these, or similar, proposals coincidentally, because by so doing the
perception would be that improvement of existing practice rather
than fundamental change would be acceptable. Understandable
though this view may be with hindsight, the fact remains that it
consolidated a public impression that the CIB and *Rethinking*

Construction were somehow in opposition, and to a certain extent duplicating each other's work.

There is, however, little indication that this 'shadowing' exercised any influence at all on the task force's approach. The widening divisions between the task force and the CIB slowly began to affect willingness among senior members of the industry giving voluntarily of their time, although it brought in many new volunteers (mostly unaware of what had been chewed over since the Latham review).

In 2000 the National Audit Office (NAO) conducted a review of the industry's performance, and of the various initiatives developed within M4I, the Achieving Excellence programme in the public sector, the Department of the Environment, Transport and the Regions (DETR, the successor to DOE) and the Office of Government Commerce. The report of the Controller and Auditor-General, under the title of *Modernising Construction*, was presented to Parliament in January 2001 (H C 87, Session 2000–2001). It confined itself to performance in the public sector, over which the NAO had a monitoring responsibility. The report was generally complimentary about the improvements emerging in the public sector's construction performance, but expressed some concern about the risk of too many initiatives going over the same ground. The report said (Executive Summary, paragraph 7):

> There are many ... organisations and networks – some privately and some publicly funded – whose aim is to promote good practice, suggesting some duplication. These have succeeded in raising awareness among the different parts of the industry – clients, contractors, consultants and specialist suppliers – but there is now a need for more co-ordination and better direction of their activities.

Discussions about how to proceed continued into July 1998, during which time the opportunity was given to all sectors of the industry to consider and build on the task force's recommendations for the adoption of demonstration projects and for the establishment of the Movement for Innovation. At Alan Crane's insistence, consideration of how best to take the ideas forward took into account the strongly held view of Sir John Egan that performance measurement had to be a cornerstone of the new industry's approach to business. Partly reflecting Egan's previous experience in the UK automotive industry (Jaguar cars), he sought advice from the Society of Motor Manufacturers and Traders (SMMT) on the benchmarking system operated within the motor industry. From the motor industry, the designers and operators of this benchmarking system, management consultants J.D. Power, indicated a willingness to use their experience in producing a similar system for the construction industry, but the scale of fees that the firm suggested was way out of the range of resources available to M4I at the time. Accordingly, having been asked directly by Nick Raynsford in August 1998, Alan Crane agreed to try to establish key performance indicators (KPIs) covering the seven clear targets set by the *Rethinking Construction* report (see Table 10.1).

Crane was able to involve representatives from all the constituent members of the CIB, as well as the full support of the Construction Directorate of DETR in the technical and statistical studies required, in a pan-industry approach to the definition and quantification of these performance indicators. The importance of clarity in presentation was fully recognised, and the 'spiders' web' chart developed; it is now widely accepted and used. Client participation in the development of KPIs was made conditional on the addition of an eighth performance indicator, covering 'client satisfaction with the product'. The official launch of the KPIs took place at a specially convened M4I conference in November 1998. Immediately prior to the conference, Construction Minister Nick Raynsford and the head of DTI's Construction Directorate, John Hobson, asked Alan Crane if he thought that the industry would in fact accept and 'run with' the performance indicator approach. He replied that the industry would not be so stupid as to refuse, and drawing on the visual concept of the spiders' web illustrated the dangers to those firms trapped on the periphery of the new approach.

Partly as a result of the success of the early work on designing and developing the KPI system, it was suggested that Alan Crane should take over the leadership of M4I. He made his acceptance conditional on the industry as a whole supporting the proposal, and on the recognition that he would have appropriate independence of operation. In particular he asked for, and obtained, the agreement of DOE that he would not be expected to constitute membership of M4I on any representational basis involving seats 'by right' for constituent bodies. He was also assured of financial support from Government amounting to £0.5 million for the first year, with the understanding that this would be matched in due course by input from the industry itself.

With this backing, an M4I team was established, with some 24 private sector firms, as well as Government departments and the local authority sector, contributing staff and services in kind. The

Table 10.1 Targets set in *Rethinking Construction*: the scope for sustained improvement

Indicator	Improvement per year	Current performance of leading clients and construction companies
Capital cost All costs excluding land and finance.	Reduce by 10%	Leading clients and their supply chains have achieved cost reductions of between 6 and 14 per cent per year in the last five years. Many are now achieving an average of 10 per cent or greater per year.
Construction time Time from client approval to practical completion.	Reduce by 10%	Leading UK clients and design and build firms in the USA are currently achieving reductions into construction time for offices, roads, stores, and houses of 10 to 15 per cent per year.
Predictability Number of projects completed on time and within budget.	Increase by 20%	Many leading clients have increased predictability by more than 20 per cent annually in recent years, and now regularly achieve predictability rates of 95 per cent or greater.
Defects Reduction in number of defects on handover.	Reduce by 20%	There is much evidence to suggest that the goal of zero defects is achievable across construction within five years. Some UK clients and US construction firms already regularly achieve zero defects on handover.
Accidents Reduction in the number of reportable accidents.	Reduce by 20%	Some leading clients and construction companies have recently achieved reductions in reportable accidents of 50 to 60 per cent in two years or less, with consequent substantial reductions in project costs.

Table 10.1 Continued

Indicator	Improvement per year	Current performance of leading clients and construction companies
Productivity Increase in value added per head.	Increase by 10%	UK construction appears to be already achieving productivity gains of 5 per cent a year. Some of the best UK and US projects demonstrate increases equivalent to 10 to 15 per cent per year.
Turnover and profits Turnover and profits of construction firms.	Increase by 10%	The best construction firms are increasing turnover and profits by 10 to 20 per cent per year, and are raising their profit margins as a proportion of turnover well above the industry average.

majority of the operating costs of the team and its administrative support were covered through organisations seconding staff for fixed periods, while continuing to absorb their salaries and supporting administration within their own firms' budget.

As noted in Chapters 8 and 9, the Construction Industry Board was keen to consolidate a strong position as the pan-industry representative body in relation to the developing influence of M4I. M4I was increasingly seen by private sector firms on both the supply and client sides as the more effective tool for the development and implementation of the innovative approach called for by the Egan task force, and this posed some sensitive questions for Crane in his capacity as M4I chairman and as chairman of the Construction Confederation, the trade association of the major contractors. His initial inclination was to step down as CC chairman, but, being keen to maintain on-going dialogue with the Board, he opted instead to remain the representative of the CC on the Board, reporting to it on matters relating to Rethinking Construction and M4I in this capacity. This somewhat convoluted relationship was not welcomed by the Board's chairman at the time, Tony Jackson, but given the perception (recorded in Chapters 8 and 9) that ministers and the Department were holding the Board at arm's length while concentrating their attention on Rethinking Construction/M4I, this seemed the best that could be expected at the time.

As the programme of work of M4I developed, particularly in

relation to the establishment of demonstration projects to illustrate best (and sometimes less than optimum) practice, consideration within the movement was increasingly given to how this could be disteminated throughout the industry. Both Alan Crane and the DOE's construction director, John Hobson, were strongly of the opinion that this should not be the responsibility of the movement itself, believing that if it was asked to do so this could only be achieved at the expense of M4I's ability to concentrate on the definition and development of the innovations for which Sir John Egan's task force had originally called.

Accordingly, the Department established a Construction Best Practice Programme (CBPP). This was launched with high profile in the Queen Elizabeth Conference Centre, Westminster, by the Minister, Nick Raynsford, supported by Christopher Wates, and David Adamson who stressed the complimentarity of each of the movement's three legs – the intellectual input of Latham, the resources of Egan/M4I, and the worked examples of the CBPP. This was not the desired political context and the press were briefed accordingly. CBPP was generously funded to the extent of some £2m. per year from the Department's budget for research and development, which it was able to site in the Building Research Establishment at Watford. The political decision had already been taken to make BRE independent of the Department on the basis of a phased programme of reductions in Government funded contracts over a period of five years, and the CBPP proved, under its Director Zara Lamont seconded from Carillion plc, an effective method of preserving the expertise of specialists to the advantage of the industry during this period. As earlier noted, a large area not appropriately covered within the Rethinking Construction/M4I movement remained private and public sector housing, and to remedy this situation the Housing Forum was established, which Sir Michael Pickard agreed to chair and which was to be included in the overall programme of Rethinking Construction/M4I. This ensured the involvement of the very important area of social housing provided by registered social landlords (RSLs), more familiarly known as housing associations, which had assumed much of the responsibility for the provision of lower-cost housing for defined social groups, including the disadvantaged, from the local authorities. They were represented collectively in the movement by the Housing Corporation, responsible for allocating and overseeing grants to individual RSLs and approving grant applications, and by the RSLs' own representative association, the National Housing Federation.

Building on the momentum resulting from the support given to

Rethinking Construction, M4I moved forward rapidly, establishing Regional Clusters around the country to involve firms prepared to make their experience available through the demonstration projects. Between July and November some 40 organisations had volunteered over 90 possible schemes for inclusion in the demonstration project programme, of which 80 per cent, with a combined capital value of over £1bn, were accepted. By the end of 2003 this had expanded to include over 440 projects with a combined capital value of over £7bn, as well as including a further 110 projects illustrating particular issues relating to the use and treatment of human resources in the industry.

Crane and many others see this development of the nationwide coverage of demonstration projects, and particularly its extension to the UK regions outside London and the South East, as one of the notable successes of the movement. Equally successful was the concentration on, and acceptance of, the principle of performance measurement, and the development of the work on performance indicators originating with the member bodies of the CIB. This led to the formulation and adoption of the key performance indicators now applied in parts of the industry. Annual reports on the performance of the industry expressed in relation to the adopted KPIs are now published and in Alan Crane's opinion this performance-based regime shows clearly that, by adopting the principles formulated as a result first of the Latham review, and subsequently of Rethinking Construction, measurable positive difference in the industry's effectiveness can be achieved quite quickly. By the end of 2003, as he points out, the indicators across the industry as a whole were on a rising curve for the first time.

Rethinking Construction/M4I undoubtedly benefited from the personal involvement of ministers and top officials. Prior to the Egan study, concern had been expressed about the difficulty of bringing the local authorities on board, given the (inevitably) prickly relationship between central and local government. Given that central and local government together account for over 40 per cent of the total demand on the industry it was vital that these sectors were seen to be involved, and Egan was quick to recognise the need to ensure appropriate 'buy-in' at the ministerial and political level. Indeed the pre-election involvement from Messrs Prescott, Brown, and Mandleson made that much easier. Before the official announcement of Sir John Egan's appointment, Nick Raynsford, Alan Crane, and Egan himself agreed to promote the setting up of a ministerial group to support the Rethinking Construction concept, which involved, among others, Geoffrey Robinson, Chief Secretary to the Treasury, Chris Smith,

Heritage Secretary, as well as DETR ministers. The idea, subsequently adopted by the Prime Minister, of design champions to influence good design in public building, originated here. A Construction Industry Forum, which brought together influential industry progress, as well as top Whitehall officials including Sir Steve Robson, at the time second permanent secretary at the Treasury and very conscious of the importance of the construction industry in facilitating the improvements promised by the Government to the public infrastructure, successfully advocated the establishment of a Local Government Task Force to contribute a local authority input to the movement.

On 'the supply side', the balance of influence had changed, albeit temporarily. The positions in the lobbying caucuses, and in the project teams advocated by Latham and CIB for designers, specialist contractors and suppliers had been weakened by Rethinking Construction/M4I's focus on the main contractor/client axis.

Alan Crane is in no doubt personally that the innovations introduced during the period of reform have had a profound effect in establishing a new mood in the industry and many would share that view. There is now no debate that supplier/customer relationships in the industry are based on the idea of integrated teams. There is also an acceptance by clients of the need to develop appropriate expertise, and to promote concepts of shared risk and reward. This is despite Crane's view that attempts to bring together a genuinely representative body of top executives of private sector clients is doomed to failure, since construction is not their main business and their concentration is inevitably focused on their own business performance. It is difficult for them automatically to recognise and accept the value added to their business by successful construction solutions. At the micro-level, managers who do recognise construction's importance to the success of their own operations see little need to approach this collectively, because they have already instituted internal procedures which they see no need, and in some cases do not wish, to make widely available. Generally speaking, such experts prefer to exchange experience in small informal and confidential groupings.

In a short time by national industry standards, a huge amount of activity had been generated, fuelled by considerable political power and financial resources. Only through analysis as the years roll by will history be able to judge if a more inclusive approach would have been as dynamic and as effective, but there can be no doubt that through Rethinking Construction/M4I, modern business realities permeated the fabric of the industry and its clients more deeply than ever before.

11 A phoenix from the ashes

Winding-down of the CIB; emergence of Constructing Excellence and Strategic Forum for Construction; development of the Clients' Charter; attempted expansion of the CCC; appointment of CCC chief executive; disintegration of client movement; collapse of CCC. Brief assessment of achievement and effectiveness of the reform initiatives as genuine catalysts for change.

The collapse of the collaborative structure for the industry, which had relied so heavily on the commitment of both the supply side elements and its customers since the Latham era, did not happen immediately. The distinct lack of enthusiasm with which the Vickers report on the CIB was greeted by both contractors and some major clients inevitably resulted in increased emphasis being given to the work of M4I and Rethinking Construction on which DETR's Construction Directorate was perceived to concentrate. Nonetheless, the CIB pushed on with a programme of work aimed at advancing policies in the priority areas which Chris Vickers had defined; the quality of resource, particularly people, the quality of product, the efficiency of operations, profitability for firms in the industry and value for money for clients.

Following the launch of the new client body, the Confederation of Construction Clients (CCC), the Confederation's new chairman, Mike Roberts, initiated discussions with the contractors' representative organisation the Construction Confederation (CC), about the line to be taken in relation to following up the Vickers' recommendations. These discussions were conducted largely within the group of chairmen and chief executive officers of the industry's umbrella bodies, known as CUB. Mike Roberts had made clear to the contractors that private sector clients, particularly the large and important clients from manufacturing, retailing, utilities, and property development would have no truck with a CIB which could be perceived as perpetuating the consensus-based approach of the old Board. Since these were precisely the type of clients the new CCC wished to bring on board there was little room for compromise. In fact, the position adopted by these clients proved very welcome to the contractors, who for their own reasons were unenthusiastic about a federated approach to the industry's affairs in which the traditional 'main contractors' would not have such a dominant position. Benefits were seen in canalising client/specialist contractor links via the main contractors. As noted previously the new CCC had refused to confirm that it would take up the offer of seats on the board of management of the successor to the CIB, and had made clear that it both could not, and in fact would not, subscribe financially to the level identified as necessary by Vickers. Accordingly when the Newbody's board of management met at the end of March 2001, the CCC's chairman stated that senior representatives of client members of the Confederation would 'continue to attend' meetings of the pan-industry body in whatever form it eventually took, but that the CCC would not contribute financially.

The CCC's chairman had discussed the position with the newly appointed M4I board chairman, Alan Crane, within the context of the launch of the CCC and of the Clients' Charter. Alan Crane wrote:

> On behalf of the M4I Board I am writing to thank you for joining us at our Awayday dinner and for sharing with us your thoughts on the work programme of the Confederation of Construction Client's and on the operation of the Client's Charter. The clarity of your presentation and your vision of the Charter as an integral part of Rethinking Construction was much appreciated by us all.
>
> I think we were all totally supportive of your view that CCC will seek to work through endorsing the innovative practices that others have developed and proved rather than seeking to develop wholly original 'models'. This is, of course, the way in which the M4I Board seeks to operate. We will ensure that those innovations in client procurement practices that emerge from our Demonstration Project programme will be shared with you.

I invited the M4I Board's views on the CCC and in particular the operation of the Client's Charter during our business session on Friday. I think you will be pleased (but not surprised) to learn that the Board was fully supportive of the ambitions of CCC and its method of working. The Board also asked me to formally confirm our full support for the objectives of the Clients' Charter. It is essential that the efforts that designers, contractors and suppliers are making to achieve sustained improvement in project performance and value for money are supported and, as you said '... pulled through' by the complementary actions of clients. Decisive actions to set targets for their own efforts and to measure improvements will, for the first time, ensure that the agenda for continuous improvement set out in Rethinking Construction is being supported by consistent action across the entire industry.

We look forward to working with you and your Supervisory Board on the on-going development of the Charter and, more generally, in ensuring that our work programmes continue to complement each other. The need to offer a service to support the procurement decisions of occasional clients is something we recognise as a priority and will wish to be associated with.

In the light of Mike Roberts' statement to the CIB board of management, the other members of the CIB, the Construction Products Association (materials and components suppliers), the Construction Confederation (contractors), and the Construction Industry Council (professionals) made clear that they would not wish to continue supporting the CIB and its successor, if the clients were not full participants. The Construction Liaison Group, representing the sub-contractors and specialists, were especially reluctant to accede to the demise of the CIB, but given their minority position, and the fact that the two main constituents in the CLG, the Specialist Engineering Contractors Group and the National Specialist Contractors Council were finding it increasingly difficult to adopt a common position in the CLG, they had little choice but to accept the inevitable. Accordingly, the chairmen and chief executives of the umbrella bodies (CUB) had prepared formal proposals for replacing the CIB with a Strategic Forum for Construction, an initiative originating with the Minister, Nick Raynsford. This Strategic Forum intended to bring together the existing interests represented in the old CIB, together with Rethinking Construction and its executive arm M4I, the newly established Commission for Architecture and the Built Environment (CABE), the Housing Forum, the Office for Government Commerce, and other bodies such as CRISP. Given the determination of those involved to concentrate on genuinely strategic issues, the British Property Federation requested its own membership of the proposed Strategic Forum, rather than through the client body, the CCC.

Meanwhile, the client movement was watching with interest the aspirations of the newly established Confederation of Construction Clients, and the launch of the Clients' Charter. The Confederation had advertised for a full-time, high-profile Chief Executive, and from a strong field, and with the strong backing of the Confederation's chairman, Mike Roberts, was able to attract Zara Lamont, a board member of Carillion plc and at the time on extended secondment from that company, as chief executive of the Construction Best Practice Programme. Zara Lamont embarked on a major programme of recruitment among public and private sector clients. Under her direction, and that of the chairman, the CCC embarked on a programme of work which concentrated on a limited number of high priority issues.

Training:

- achieving a fully badged Construction Skills Certification Scheme (CSCS) workforce of competent operatives;
- managing training for the supply side;
- developing good client skills.

Measurement:

- having a preferred, consistent and continuous system of measurement of the effectiveness of all projects.

Briefing:

- improving the whole process of client briefing.

Integrated supply chain:

- guidance to clients on creating an integrated supply chain and construction teams.

Selection criteria:

- developing a consistent approach to selection criteria for choosing supply organisations, e.g. safety, health, equal opportunities, environmental sustainability aspects, integrated teams.

The Clients' Charter was seen as, and remains, a key component in the strategy of improving clients' skills and their relationships with their suppliers. It offered clients the opportunity to establish a new culture through a structured programme of change, supported by measurement, and the exchange of best practice experience. It required a commitment continually to improve client performance, and monitor and assess their own progress in achieving targeted levels of improvement, measured against key performance indicators (KPIs) adopted and promulgated through Rethinking Construction. To assist the CCC in operating and developing the charter, organisations specialising in information management and systems were invited to tender for the project, and a contract was awarded to Achilles Information Ltd, working in collaboration with Birmingham University.

The charter has proved to be one of the lasting achievements of this period, and could indeed have provided a very effective tool for implementing the improvements in client skills which had been identified and recommended by the CIB, and subsequently developed in the light of Rethinking Construction. Unfortunately, however, a number of the client organisations which had accepted membership of the new CCC on the condition that they both subscribed to the principles of the Clients' Charter, and implemented its procedures in their own processes, found themselves unable publicly to deliver on this undertaking. Some key clients were unhappy with the procurement philosophy which the CCC chairman had adopted in his time at BAA, which they interpreted as insufficiently collaborative. Compromise within the deliberations of the CCC proved impossible. Member firms of the BPF were particularly determined in their view that what might appear to be external pressures to change commercial attitudes in their dealings with their suppliers should be resisted, since their performance as reflected on the stock market suggested that they were doing very well with the 'status quo'. Nonetheless, the charter concept developed in other client sectors, notably in the social housing sector, where the Housing Corporation insisted on its application by individual housing associations as a condition of receiving Housing Corporation approved grants.

The second quarter of 2001 saw the implementation of the decision to replace the Construction Industry Board by the new Strategic Forum for Construction, and the Minister, Nick Raynsford, asked Sir John Egan to become the first chairman with effect from 1 July 2001. The Minister had taken this initiative in his capacity as sponsoring minister of the industry, and as president of the former

CIB, on the advice of his officials, who were firmly convinced that personal intervention by the Minister was required to clear the impasse that had developed as a result of the demise of the CIB and the lack of genuine support for the new CCC. The programme of work that developed in the Strategic Forum under Sir John's chairmanship has been recorded in the previous chapter, but one of the major preoccupations of the time that should be emphasised was the need radically to improve the industry's record in health and safety. A series of high-profile site accidents, in excess of the national norms for industry as a whole, had resulted in the Health and Safety Executive calling for major improvements. The industry, to its credit, responded with a collective determination to seek to improve matters, and with the combined influence of the Department, Sir John Egan, and government departments collectively, an implementable action plan emerged. The CCC committed itself to a policy of making award of contracts conditional on evidence that suppliers, whether main or sub-contractors, insisted on their staff possessing recognised skills certification, and this with the full support of the Construction Industry Training Board (whose chairmen, Hugh Try and then Sir Michael Latham, greatly raised the profile and success of CITB) gave a major boost to schemes such as the Construction Skills Certification Scheme (CSCS), which now has nationwide application.

In May/June 2001 another general election took place, and the Labour Government was returned with a large majority. Immediately changes to the machinery of Government were made. There was no longer to be a minister for the industry, and many saw that as a demotion of the industry in Government priorities. Responsibility for the Government's relationships with, and sponsorship of, the construction industry was transferred from Environment to the Department of Trade and Industry. The reasons for this transfer remain obscure, although in later conversation, John Hobson, the head of the Construction Directorate at the time, maintained that it was as a direct result of expressed preference by the industry. There seems little objective evidence of this, and indeed the industry's unique relationship with a government department that originated from the wartime Ministry of Supply, through the Ministry of Public Building and Works, and eventually Environment had in fact assured it of more direct attention on the part of Government than it could expect as just one, albeit important, national industry among all others overseen by DTI.

Graham Watts, chief executive of the CIC, has pointed out in a comment to the authors that several left-wing think tanks, as well as some leading journalists, had made the claim that the industry should not be sponsored by the same department of Government that regulated it, and that there were significant conflicts of interest in this dual role. There were, says Graham Watts, several damaging claims to this effect in George Monbiot's book *Capture State*, published about this time. The Construction Products Association (CPA) had, according to Watts, expressed a preference for a move to the DTI, which was adduced as evidence of industry support.

The new Minister covering construction, inter alia, was Brian Wilson MP, with the official title of Minister of State for Energy and Industry, and he rapidly re-emphasised the support that the Government intended to give to promote the success of the new Strategic Forum. New officials, Elizabeth Whatmore as director and Rodger Evans as having direct responsibility for the Government's relationship with the Strategic Forum, were appointed, and the new directorate guaranteed secretarial and funding support. Responsibility for health and safety remained with Nick Raynsford, who transferred to the new Department of Transport, Local Government and the Regions as Minister for Local Government, as did responsibility for the building regulations. This allocation of responsibilities recreated the rather fragmentary approach to the industry's operations and its regulation which the comprehensive coverage of the former DOE and DETR had successfully avoided.

Throughout the rest of the year 2001 and the early months of 2002, John Egan, through the Strategic Forum, and particularly through the Rethinking Construction Movement (see Chapter 10) which had appointed David Crewe as its director, pressed for speedy implementation of the approach to measurement and improved quality which had featured so prominently in his first report. The follow-up, *Accelerating Change*, was published in September 2002, and reviewed progress since *Rethinking Construction* had first appeared in 1998.

At the same time, and with the full support of the industry through the Strategic Forum, M4I, and the industry sector bodies, the Minister asked Sir John Fairclough to review the role that Government should play in supporting construction research. Public funding for construction research had been between £50m. and £70m. annually for the previous ten years. The former DETR Programme (now DTI and DTLR) and that of the Engineering and Physical Sciences Research Council (EPSRC) provided the bulk of it. Sir John's report concentrated on funding provided by DTI and

DTLR, which had a combined budget of around £22m. for supporting construction innovation and research.

One of the main catalysts for the review was the ending of the
Government's framework arrangement with the Building Research
Establishment (BRE) in March 2002. BRE was until 1997 a part of
the old Department of the Environment, undertaking research work
to support the department's regulatory and sponsorship role. In 1997
it was privatised, with a five-year guarantee of a minimum amount of
work which would be offered on an exclusive tender basis by the
Department.

Fairclough's main recommendations were:

- Industry, led by the Strategic Forum, should set out how construction can contribute to the quality of life agenda. The strategy would provide a framework for future planning and
  investment in education and skills, capital infrastructure, and
  research and development (R&D).
- Government should safeguard investment in construction R&D
  and better target investment in productivity, ensuring value for
  public sector clients and strategic issues.
- The skills problem in construction, and its impact on R&D, must
  be addressed. Attempts must be made to make programmes of
  work exciting for researchers by defining work in quality of life
  and sustainability terms rather than as narrow construction problems. A high profile generalist construction qualification should
  be developed to encourage the best young talent into construction. Government should demand multidisciplinary teams and
  more interchange of people between industry and the academic
  world.
- Government should commission longer-term programmes of
  R&D, on merit, encouraging collaboration to ensure relevance
  to industry needs.
- Specialist research teams should no longer be maintained by
  Government 'just in case' but centres and networks of excellence
  should be developed and encouraged to take responsibility for
  wider sector issues.
- The traditional construction research organisations should be
  encouraged to work more closely together and learn from best
  practice and innovation abroad and in other sectors.

The concern with utilising the expertise of the research world to
advance the improvements constantly being advocated had been

highlighted by the Prime Minister, Tony Blair, who had commissioned and written a Foreword for a report, *Better Public Buildings*, which sought, with the aid of CABE, to make the public aware of good design in construction. He wrote:

> The UK needs to raise its game in the provision of public buildings and infrastructure. Delivery of the Government's huge programme of infrastructure investment requires an effective construction industry delivering good value for money.

It was clear throughout 2001 and 2002 that much of the work in the periods since Latham and Egan reported was beginning to bear fruit, at least in the increasing public awareness of the importance of construction, and in the real and often successful attempts to bring together and implement the various policy areas relating to procedures, best practice, research and Government involvement. There was also evidence, through the charter and elsewhere, that clients were anxious to improve, and that this client awareness was spreading throughout the public and private sectors. It might have been expected therefore that the newly formed client movement would prosper and become a genuine influence on the industry's willingness to implement the detailed improvements recommended. A move was made to strengthen the client representation by merging the newly formed CCC with the Construction Round Table, and it was hoped that this would prove the catalyst necessary to bring on board the major private sector clients who had so far stood aloof from the collective activity in the industry. In fact the reverse proved to be the case. The high public profile adopted by the CCC's chief executive, Zara Lamont, kept the concept of the client's leadership of the industry well in the public eye. This ensured that, as initiatives and proposals for action emerged from the Strategic Forum and from the peripheral organisations that Rethinking Construction sought to involve, there was always a strong and united call from the industry for full client participation. Throughout 2001 and the first half of 2002, no fewer than 90 separate presentations on behalf of clients were made to different construction-related organisations in the UK. This major input of intellectual resource was, however, not matched at all by a commensurate rise in membership of the client organisation. Although by the start of 2002 Zara Lamont had succeeded in bringing into the Confederation representation from central and local government, the utilities, one of the universities, the voluntary housing sector and some private sector organisations through the merger with the Construction Round Table, the major retailers, property developers, manufacturers and distributors, many with large programmes of construction, still proved elusive. Even among the new member organisations, a number found themselves unable to meet their obligations of funding support for the Confederation, and of releasing senior experts in their organisation to direct and advance its programme of work. The Confederation had embarked on an ambitious programme, including a commitment to produce training resources for small and occasional clients, which the Government was anxious to promote, since over 80 per cent of contracts were awarded by clients in this category.

Always poorly funded, by the start of 2002 the Confederation's supervisory board was warned by their officers that funding was insufficient to ensure the organisation's viability for longer than a further year. The Confederation had contributed in a major way financially to the development of the charter, and as evidence of the importance that the Government continued to give to the development of client skills, the DTI gave financial cover to the CCC to allow this to continue. The Confederation's supervisory board decided by June 2002 that they would not be justified in continuing to work in the same way, and, bearing in mind that the Confederation had become registered as a private company limited by guarantee, in order to allow it to own the charter and contract with Achilles Information Ltd to develop it on a commercial basis, decided to put the CCC into liquidation by the end of 2002, just two years after its painful birth. Ownership of the charter was transferred to Constructing Excellence, a body established within the Rethinking Construction movement which incorporated M4I and CBPP.

It might appear as though this rather inglorious ending of the Confederation, and of the CIB earlier, reflected a failure of the movement started with such high hopes by Sir Michael Latham in 1994. In fact this has proved not to be so. By 2004 the Strategic Forum was well established, and, to all intents and purposes, has embraced and built on the concept of the Construction Industry Board. A Construction Clients' Group, administered by the BPF, which is very similar to the original Construction Clients' Forum (CCF) has replaced the former Confederation of Construction Clients, and appears to have involved the CBI and Institute of Directors, as well as the public sector representative bodies which were always the prime movers in seeking the establishment of a genuine and powerful client movement. Constructing Excellence seems to have the financial and intellectual resources that were denied to the

Construction Industry Board, and is making available practical and implementable guidance to the industry at large.

In 2003 Adrian Barrick, editor of *Building* magazine, wrote a leader entitled 'Our go', in which, very much with tongue in cheek, he called for another industry review. He wrote:

> Time for another industry review, don't you think? After all, it's 10 years since Latham and five since Egan. The next one ought to be spearheaded by the Strategic Forum, but since that exalted organisation has vanished – at the instigation of the contractors – we thought we'd have a go. Our review didn't take a year (we did it over lunch), and it didn't involve politicians (we thought some construction people might be more helpful). We're sorry we couldn't come up with a portentous title like Rethinking Change or Accelerating Construction. But we didn't think you'd mind.
>
> If you look beyond the Pythonesque illustrations, we do have some serious points – particularly the need for lemon marmalade in site canteens. Like many of our proposals, it's aimed well below board level (in this case, at the catering manager). That also goes for better programming, fewer drawings, fewer emails, and the scrapping of *Building*'s legal pages. Which is down to me, I guess. (I might have to think about that one.) After the 'top down' approach of Latham and Egan, the essence of our report, then, is 'bottom up' change. If New Labour were in charge, they'd call it The People's Review. Thank goodness they aren't.

But he went on:

> We must admit we've been lucky with timing. Construction is in much better shape now than when either Latham or Egan conducted their studies, so we've no need to say things like 'if only building could be more like the car industry – apart from British Leyland'. And it would be churlish not to pay tribute to our predecessors. Latham persuaded the team to stop fighting each other and deliver a better service to clients. Egan did much to unite the industry (in opposition to himself). But, in fairness, he did convince ministers to reform the Government's antediluvian procurement regime. Even civil servants can now, like property developers, use phrases such as 'partnering' and 'supply-chain integration' while screwing tender prices to the floor.

And this surely is the point. We may have come full circle, back to the approach to reform of a decade ago. But the fact is that any genuinely objective assessment of what has happened in the years 1993–2002 shows convincingly that this movement for reform has been more influential than any in the years since the war. The costs, in time given voluntarily – probably over two million hours – and in tax-payers' money, have been huge: by no means everything, or perhaps even the greater part, of what was hoped for has been achieved, but those who have been involved, and who have contributed so much intellectual and practical input to the improvement of the industry, would be justified in taking pride in their efforts and in the demonstrable successes. Latham provided the intellectual basis for the reform movement, Egan obtained the resources necessary for the dynamic approach he favoured, the Government gave unstinting support, and it is now up to the industry to ensure that it fully reaps the benefits. This huge industry is better now, and it knows it is in need of more improvement. There is more emphasis on productivity and fairness than there was, and the rising dominance of litigation has been halted and reversed. People in this industry are becoming better instructed, better skilled, better treated, and more productive. They have a better chance to enjoy their work. The extent of this increased potential for improvement is the content of the second part of this report, which has tried to demonstrate how enormous, and how political, the reform has been over this historic decade. It has concluded that the game has been worth the candle.

L'envoi

It is now necessary to draw lessons from the experiences of the last decade, and to investigate how this increased potential for improvement of the industry can be effectively implemented. This is to be the subject of further work supported by the Foundation for the Built Environment and BE, and covers essentially two main elements. The first is to provide collated evidence, which is now coming forward, of the measured advantages resulting from the application of the new processes which have evolved from the collective work of the industry. This involves analysing specifically how these improvements impinge on the performance of organisations in both supply and client sides, resulting in enhanced long-term health for such businesses. Organisations need to have convincing figures showing the enhanced profitability that will accrue to them through the adoption of the new approach to construction which is described here. Boards of directors need to recognise the factors that may inhibit adoption of the defined innovations developed over the last decade, particularly in relation to integrated management and risk acceptance. To be convinced they need to see reputable figures illustrating the difference in value for money between projects concluded before and after adoption of the new methods, and successful clients and suppliers are providing such evidence with sufficient clarity, although it is difficult to assess success quantitively.

The second area of learning from experience relates to how such comprehensive attempts by an industry and by Government to identify the need for, and then implement, fundamental industrial change can best be undertaken. It is this area which should prove of most interest to future business and management school students, the future directors and innovators of the national industry. Human relationships, the dynamics of organisations, and government/industry and government/media relations, as well as the best uses of academic and research resources need to be defined and assessed in order to ensure that the benefits of all the effort expended are maximised.

Postscript

Despite the initiative fatigue at the end of the decade of reform, and the apparent loss of momentum following the collapse of the combined client movement in 2002/2003, the subsequent two years have in fact seen wider acceptance in the industry and, throughout the public and private sectors of the UK economy, of the co-operative and more integrated approaches recommended in the Latham and Egan reports. This in turn has produced perceived, measurable and audited improvement in public and private sector construction programmes. High profile failures in certain prestige projects, in relation to cost and time overruns, design failures and technical shortcomings, have of course made media headlines given the preference of the UK media for failure, but the Government's and industry's commitment to achieving the improvements recommended by Latham and Egan has ensured recognition of the importance of construction to the national economic well-being, and the need to preserve and develop this through collaboration rather than contractually.

Notwithstanding political downgrading of the level of ministerial responsibility for the industry, the original attempts to achieve a permanent, standing policy making body truly representative of all participants in the construction process and the development of the industry through the Construction Industry Board, although not in themselves sustainable over the change in Government, led, after considerable argument, false starts, and politically partisan posturing, to the establishment of a Strategic Forum for Construction. This forum has proved able to involve leading personalities of the industry in defining and promulgating the strategies that the industry should adopt to achieve targeted improvements based on the two reports' recommendations. The current influence of the Strategic Forum is not to be under-estimated, given the major programme of construction over the next decade contingent on the Labour Government's declared intentions for public sector infrastructure investment. The confidence in its position in the industry, and the support it enjoys among the constituent parts of the industry, have allowed the Strategic Forum to issue codes of recommended practice which have been widely accepted and implemented. The latest of these, for example, is a 'Construction Commitment' for the industry to set out its commitment to the preparation of the 2012 Olympics so that the event is well-produced, well-procured and a credit to the nation. It is interesting that the structure of the Strategic Forum from January 2006 is very similar to that of the original Construction Industry Board.

In terms of the development of good practice in the industry, the expertise originally deployed in the various technical and professional groups set up in collaboration with the first Construction Industry Board has spawned the Constructing Excellence movement as a source of advice, performance measurement, and best practice. In tune with the original vision of Latham and Egan this movement now formally incorporates the input of a representative client body, the Construction Client Group. Although this new client body is but little different in its objectives and methods of working from the original Construction Clients' Forum (CCF), and its successor the Confederation of Construction Clients, it has been able to incorporate a wider and more representative public and private sector membership than the earlier bodies, free from the internal bickering and organisational and bureaucratic pre-occupations that so adversely affected them in the latter stages after initial successes of the CCF. To add to further strength, the powerful research, analysis and best practice body, Be – as in 'Be Excellent' – has merged with Constructing Excellence.

Meanwhile, the success of the CITB-Sector Skills Council in increasing the capacity, safety, and attractiveness of the industry, has increased. There has not been as much progress in consolidating and making more pervasive the benefits of the often excellent research in the industry.

In the important area of design, the Commission for Architecture and the Built Environment (CABE) has, with widespread support, established a central position, in both the intellectual and policy formulating movements within UK society, with a role that some expected the Royal Institute of British Architects to take. Given the concentration of the Labour Government on increasing the provision of housing to meet changing demographic and social needs, the collaboration of the Strategic Forum, Constructing Excellence and CABE with the Housing Forum is further evidence of the increasing recognition of the needs of construction in the development of policy at the national and local level. CABE's recognition of the need for clients to be more aware of how their construction solutions can be met, and then later, how well they have been met, to the advantage of all, has led it to issue a clients' guide – *Creating Excellent Buildings* – which, where understood and implemented, is already starting to create in the UK an appreciation of the advantages of good design in construction, comparable to that already evident in other EU countries and in the United States. In central government, the Office of Government Commerce (OGC) has also increased focus on good sustainable design to achieve whole-life value for tax-payers' money, better gearing of the capacity of the supply side of the industry to the increased demand, and the wider and deeper acceptance of better ways of procurement.

So we now begin to see the effect of this wider acceptance of the vision of the Latham and Egan reports, as developed by the various movements set up collaboratively over the last decade and manifested in measurable improvements in performance and industry output. Many routine and prestige projects (such as Heathrow's Terminal 5, and commercial developments in London and regional centres) have publicised their project management's commitment to the collaborative approach, and their control of budgets and timetables evidence their successes. One of the encouraging signposts to the emerging success of the movement must be the conclusion of the March 2005 National Audit Office (NAO) report of the review of progress in the central government part of the public sector since 2001, in achieving the strategic targets set out in the Achieving Excellence in Construction programme, promulgated by OGC and Constructing Excellence. The NAO recorded that by March 2005, 55 per cent of projects initiated by central government departments were delivered to budget compared with 25 per cent in 1999, and that in the same period 63 per cent of projects were delivered to time compared with 34 per cent in 1999. (Had these improvements not been achieved, and construction practice in central government continued as before, there would have been an estimated extra expenditure on public sector projects managed by Government departments in excess of £77 million on just a sample of £1.2bn of Government projects from April 2003 to December 2004, hence a 6.5 per cent improvement over the 20 months.) There are no grounds for complacency, and there are far too many too highly publicised failures in the industry, but it would be churlish not to recognise that the collaborative efforts of the industry in reforming itself have begun to bear fruit. Since the end of its decade of reform, the construction industry has become more productive, with more capacity, although without widespread reduction in out-turn costs. In fact, for most clients, professional and domestic, the inflation due to supply pressures has exceeded productivity gains. However, overall profitability has increased a little, and that has had beneficial effect in the regard of the City for the industry. Also, as Sir Michael Latham predicted, his review and the Act have brought down the level of litigation in the industry, and this has reduced overhead costs, and the misery and loss of motivation associated with litigation. There have been slow but sustained increases in fairness, self-respect, and job-satisfaction in the industry. All of this has been stimulated by the necessarily complex, exasperating, stimulating, and moderately successful decade of reform.

Appendix I

Summary of recommendations of the report *Constructing the Team* by Sir Michael Latham

Reprinted from the Final Report of the Government/Industry review of procurement and contractual arrangements in the UK construction industry (2004). References are to the appropriate chapters and paragraphs of the report.

1 Previous reports on the construction industry have either been implemented incompletely, or the problems have persisted. The opportunity which exists now must not be missed (Chapter 1, paragraph 1.10).

2 Implementation begins with clients. The Department of the Environment should be designated by ministers as lead Department for implementing any recommendations of the Report which ministers accept. Government should commit itself to being a best practice client. Private clients have a leading role, and should come together in a Construction Clients' Forum. Clients, and especially Government, continue to have a role in promoting excellence in design (Chapter 1, paragraphs 1.17–19).

3 The state of the wider economy remains crucial to the industry. Many of the problems described in the Interim Report, and also addressed in this Final Report, are made more serious by economic difficulties. But others are inherent (Chapter 2).

4 Preparing the project and contract strategies and the brief requires patience and practical advice. The CIC should issue a guide to briefing for clients (Chapter 3, paragraph 3.13). The DOE should publish a simply worded Construction Strategy Code of Practice (Chapter 3, paragraphs 3.14–3.15) which should also deal with project management and tendering issues (Chapter 6).

5 The process plant industry should be consulted by the DOE, and be part of the Construction Clients' Forum (Chapter 3, paragraph 3.18).

6 A check list of design responsibilities should be prepared (Chapter 4, paragraph 4.6).

7 Use of Co-ordinated Project Information should be a contractual requirement (Chapter 4, paragraph 4.13).

8 Design responsibilities in building services engineering should be clearly defined (Chapter 4, paragraph 4.21).

9 Endlessly refining existing conditions of contract will not solve adversarial problems. A set of basic principles is required on which modern contracts can be based. A complete family of interlocking documents is also required. The New Engineering Contract (NEC) fulfils many of these principles and requirements, but changes to it are desirable and the matrix is not yet complete. If clients wish, it would also be possible to amend the Standard JCT and ICE Forms to take account of the principles (Chapter 5, paragraphs 5.18–5.21).

10 The structures of the JCT and the CCSJC need substantial change (Chapter 5, paragraphs 5.26–5.29 and Appendix IV).

11 Public and private sector clients should begin to use the NEC, and phase out 'bespoke' documents (Chapter 5, paragraph 5.30). A target should be set of $\frac{1}{3}$ of Government funded projects started over the next four years to use the NEC.

12 There should be a register of consultants kept by the DOE, for public sector work. Firms wishing to undertake public sector work should be on it (Chapter 6, paragraph 6.11).

13 A DOE-led task force should endorse one of the several quality and price assessment mechanisms already available for choosing consultants (Chapter 6, paragraph 6.11).

14 The role and duties of Project Managers need clearer definition. Government project sponsors should have sufficient expertise to fulfil their roles effectively (Chapter 6, paragraph 6.18).

15 A list of contractors and subcontractors seeking public sector work should be maintained by the DOE. It should develop into a quality register of approved firms (Chapter 6, paragraph 6.24). The proposed industry accreditation scheme for operatives should also be supported by the DOE (Chapter 7, paragraph 7.10).

16 Tender list arrangements should be rationalised, and clear guidance issued (Chapter 6, paragraph 6.32). Advice should also be issued on partnering arrangements (paragraph 6.47).

17 Tenders should be evaluated by clients on quality as well as price. The NJCC recommendations on periods allowed for tendering should be followed (Chapter 6, paragraph 6.39).

18 A joint Code of Practice for the Selection of Subcontractors should be drawn up which should include commitments to short tender lists, fair tendering procedures and teamwork on site (Chapter 6, paragraph 6.41).

19 Recent proposals relating to the work of the Construction Industry Training Board (CITB) need urgent examination (Chapter 7, paragraphs 7.16–7.18).

20 The industry should implement recommendations which it previously formulated to improve its public image. Equal opportunities in the industry also require urgent attention (Chapter 7, paragraph 7.23).

21 The CIC is best suited to co-ordinate implementation of already published recommendations on professional education (Chapter 7, paragraph 7.30).

22 Existing research initiatives should be co-ordinated and should involve clients. A new research and information initiative should be launched, funded by a levy on insurance premia (Chapter 7, paragraph 7.40).

23 More evidence is needed of the specific effects of BS 5750 within the construction process (Chapter 7, paragraph 7.46).

24 A productivity target of 30 per cent real cost reduction by the year 2000 should be launched (Chapter 7, paragraph 7.48).

25 A Construction Contracts Bill should be introduced to give statutory backing to the newly amended Standard Forms, including the NEC. Some specific unfair contract clauses should be outlawed (Chapter 8, paragraphs 8.9–8.11).

26 Adjudication should be the normal method of dispute resolution (Chapter 9, paragraph 9.14).

27 Mandatory trust funds for payment should be established for construction work governed by formal conditions of contract. The British Eagle judgement should be reversed (Chapter 10, paragraph 10.18).

28 The Construction Contracts Bill should implement the majority recommendations of the working party on construction liability law (Chapter 11, paragraph 11.15).

29 'BUILD' insurance should become compulsory for new commercial, industrial and retail building work, subject to a de minimus provision (Chapter 11, paragraph 11.24).

30 An Implementation Forum should monitor progress and should consider whether a new Development Agency should be created to drive productivity improvements and encourage teamwork. Priorities and timescales for action are suggested (Chapter 12).

Appendix II

Key points of the 1989 report *Building Towards 2001* prepared by the National Contractors' Group of the Building Employers' Confederation and the Centre for Strategic Studies in Construction at the University of Reading

Education and training	Organisation and structure	Research and development	Image
In construction, manpower is fragmented at every level – consultants, contractors, and manufacturers operate often quite independently. This leads to misunderstandings, a lack of comprehension of others' problems and in-built suspicion and contributes to the industry's poor performance.	The traditional industry structure defines different goals for the parties within the construction team, some of which interfere with the balance between project cost, completion time, and quality.	Dramatic improvements in productivity and quality, coupled with improved management of an integrated design and construction process, are needed to bring UK construction practices up to the level of the best in the world.	The construction industry's image is considered to be poor. Accordingly, it must be improved dramatically if it is to attract credibility as an exciting, challenging, and responsible industry.
At least five professional institutions control the accreditation of degree courses in universities,	Adversarial roles are part and parcel of the traditional contract structure, with confrontation often leading to greater reward than might arise from pursuing the objective of economical construction of the required quality.	Changes are necessary to ensure that design takes account of productivity and quality issues at every stage.	The industry needs to work collectively towards identifying and publicising its success stories.

Television programmes, videos, quality books on |

continued

Education and training	Organisation and structure	Research and development	Image
polytechnics, and colleges, yet they remain fairly isolated with regard to their membership and learned society activities.	A new form of contract is needed that arranges the parties to a construction project in such a way that they all have identical goals of a timely, economical, profitable, and high quality product.	Contractor-sponsored R&D. A National Construction Productivity Centre is proposed to identify methods by which production and productivity can be improved and to promote a productivity culture within the industry.	building, a nationwide schools initiative, and collaborative programmes with schools of architecture are all needed to help enhance the industry's image and attract the best people.
Universities, polytechnics, and colleges run a range of courses in architecture, building engineering, building services engineering, surveying, and building that have evolved largely in an ad hoc manner.	People in construction need to have their energies channelled away from playing contractual games towards finding ways of creating client satisfaction.	Studies to improve quality are essential. A system for determining the appropriate quality for construction embracing total quality management must be initiated.	Image can only successfully reflect the reality that exists. Therefore, this can only be matched by an improvement in the industry's perceived performance.
A Council for Construction Education and Training, driven by senior representatives from across the industry, is needed to manage effort and to act as a single voice to government.	There are two distinct phases of a contract – concept and delivery. The conceptual design work needs to be separated both organisationally and contractually from the production process.	Studies to improve management, especially in communication across interfaces, and the training of supervisors are needed.	
A network of Centres for the Built Environment across the country involving colleges, polytechnics, and universities should be created to help break down fragmentation.		Productivity is 10 to 20 per cent higher in Germany and France.	

Education and training	Organisation and structure	Research and development	Image
There should be a common first degree that enables people to enter the industry and to specialise via a masters degree at a later stage in a particular discipline. At this stage, the professional institutions should take over, as in the medical profession, and supervise the masters degree.		Value engineering concepts should be included in contracts, production analysis should be applied to designs and total quality management needs to be practised.	
Engineering degrees are too narrow, too academic and need to include more project work. Architecture degrees are not functional or technical enough. Both degree courses should address technical, social, and economic issues.			
There is a fundamental imbalance between the demand and supply of the various construction disciplines. Only 20 per cent of the industry is civil			

continued

Education and training	Organisation and structure	Research and development	Image
engineering yet almost half of all students graduate from civil engineering courses. While 80 per cent of construction work is building, only 10 per cent of the industry's graduates are trained in building.			
The terms architect-engineer or architect-constructor should be adopted to enhance the status of engineers and builders.			

Index

eBooks

eBooks – at www.eBookstore.tandf.co.uk

A library at your fingertips!

eBooks are electronic versions of printed books. You can store them on your PC/laptop or browse them online.

They have advantages for anyone needing rapid access to a wide variety of published, copyright information.

eBooks can help your research by enabling you to bookmark chapters, annotate text and use instant searches to find specific words or phrases. Several eBook files would fit on even a small laptop or PDA.

NEW: Save money by eSubscribing: cheap, online access to any eBook for as long as you need it.

Annual subscription packages

We now offer special low-cost bulk subscriptions to packages of eBooks in certain subject areas. These are available to libraries or to individuals.

For more information please contact webmaster.ebooks@tandf.co.uk

We're continually developing the eBook concept, so keep up to date by visiting the website.

www.eBookstore.tandf.co.uk

Printed in the United States
by Baker & Taylor Publisher Services